Rainfall - Extremes, Distribution and Properties

Edited by John Abbot and Andrew Hammond

Published in London, United Kingdom

IntechOpen

Supporting open minds since 2005

Rainfall - Extremes, Distribution and Properties
http://dx.doi.org/10.5772/intechopen.77580
Edited by John Abbot and Andrew Hammond

Contributors
Ana Barros, Yajuan Duan, Prabhakar Shrestha, Steven Paul Chavez, Chang Jin Ma, Gong-Unn Kang, Ahmed Youssef, Mazen Mustafa Abu Abdulla, Fawzy Nashar, Emad Abu AlFadail, Yuriy Kuleshov, Msafiri Mkonda, John Abbot

First published in London, United Kingdom, 2019 by IntechOpen
IntechOpen is the global imprint of INTECHOPEN LIMITED, registered in England and Wales, registration number: 11086078, 7th floor, 10 Lower Thames Street, London, EC3R 6AF, United Kingdom
Printed in Croatia

British Library Cataloguing-in-Publication Data
A catalogue record for this book is available from the British Library

Additional hard and PDF copies can be obtained from orders@intechopen.com

Rainfall - Extremes, Distribution and Properties
Edited by John Abbot and Andrew Hammond
p. cm.
Print ISBN 978-1-78984-734-5
Online ISBN 978-1-78984-735-2
eBook (PDF) ISBN 978-1-83962-803-0

For EU product safety concerns:
IN TECH d.o.o., Prolaz Marije Krucifikse Kozulić 3, 51000 Rijeka, Croatia, info@intechopen.com or visit our website at intechopen.com.

Meet the editors

Dr John Abbot is currently a Senior Research Fellow at the Institute of Public Affairs in Australia, and has an adjunct research position at the University of Tasmania. He has worked as a research scientist in industry and universities including Queens University (Canada), the University of Tasmania, and Central Queensland University. Dr Abbot has a BSc (Imperial College), MSc (University of British Columbia), MBiotech (University of Queensland) and a PhD (McGill University). He has published over 100 articles, including 12 on the application of neural networks to climatic phenomena, particularly rainfall forecasting. Interest in rainfall forecasting was stimulated by the severe flooding in Queensland during 2010/2011, when the low skill of official rainfall forecasts using general circulation models became evident.

Dr Andrew Hammond is a Senior Lecturer in Geoscience within CQUniversity's (formerly Central Queensland University) School of Engineering and Technology at its Rockhampton North Campus in central Queensland, Australia, which is located close to the southern portion of the World Heritage listed Great Barrier Reef. He has gained BSc (Hons) degrees in geology and geography from the University of Tasmania, Australia; a MApplSc in Soil Science from Lincoln University, New Zealand and a PhD in Earth Science from Massey University, New Zealand.
Dr Hammond and his research students' focus is on geohazards (mass movement and floods), pedology, hydrogeology of coastal catchments, geomorphology, sedimentology and mine geology within central Queensland.

Contents

Preface

This book describes aspects of rainfall including the extremes, distribution and properties. The introductory chapter in the first section focusses on drought and flooding rains over Australia, placing extreme rainfall events from recent decades into a historical context. Although recent droughts and floods may often be perceived as unprecedented, when the extended instrumental records are considered, and also proxy reconstructions extend back over hundreds of years, this shows these are recurrent events in the natural climate cycles.

The three chapters in the second section focus on distribution and impacts of rainfall in different parts of the world. The first chapter presents a statistical analysis of rainfall patterns for Jeddah City in Saudi Arabia and considers future impacts. Jeddah has been hit by many rainstorm events during the past decade, which have increased dramatically since 2009, representing changing rainstorm patterns. The study indicated that Jeddah is characterized by two patterns of rainfall: one for rainfall values below 50 mm and the other for values above 50 mm. The second chapter in this section examines rainfall in the context of impacts, vulnerability and climate change in eastern Africa, estimating the causes, vulnerability, impact and adaptation strategies associated with climate change.

The third chapter in Section 2 focusses on a demonstration project initiated by the World Meteorological Organization using space-based weather and climate extremes monitoring. This was established to run for two years (2018-2019) focussing on weather and climate extremes, particularly drought and heavy precipitation over South-East Asia and the Pacific. Case studies presented illustrate drought monitoring in Australia in 2007 and 2018, and also heavy precipitation in Australia in 2010, and Thailand and Malaysia in 2014, which caused widespread flooding.

The third section comprises two chapters that focus on physical and chemical properties of rainfall. One chapter examines the role of aerosols that act as cloud condensation and ice nuclei. These phenomena play critical roles in cloud thermodynamics, and cloud and precipitation microphysics, known as the aerosol indirect effect. In the study presented, aerosol-cloud-precipitation interactions in the Central Himalayas were examined using aerosol observations to specify cloud condensation activation properties for simulations of a pre-monsoon convective storm.

In the final chapter, the relationships between pollutant scavenging effect and the size of raindrops was investigated. Raindrops were collected in Fukuoka City, Japan, and segregated as a function of their size using a specifically designed raindrop collector. The chemical nature of individual size-resolved raindrops was determined. It was found that the elements S, Ca, Si, and Al had relatively high concentrations in raindrops, especially small raindrops. Most of the elements measured showed a continuous decrease in concentration with increasing raindrop diameter, and it was concluded that rainfall plays a valuable role in scavenging natural as well as artificial particles from the atmosphere.

John Abbot
Institute of Public Affairs,
Melbourne, Victoria, Australia

University of Tasmania,
Tasmania, Australia

Climate Modelling Laboratory,
Noosa, Queensland, Australia

Section 1

Introduction

Chapter 1

Introductory Chapter: Australia—A Land of Drought and Flooding Rain

John Abbot

1. Introduction

Instrumental records of Australia's rainfall are maintained by the Australian Bureau of Meteorology (BOM) and extend back over 150 years in some locations. Annual contour maps of the continent extending back to 1900 illustrate how annual rainfall has varied across the continent [1]. The continent has a long history of droughts and floods [2], extending into recent years [3, 4]. **Figure 1** shows a map of Australia with the individual states and various locations referred to in this chapter.

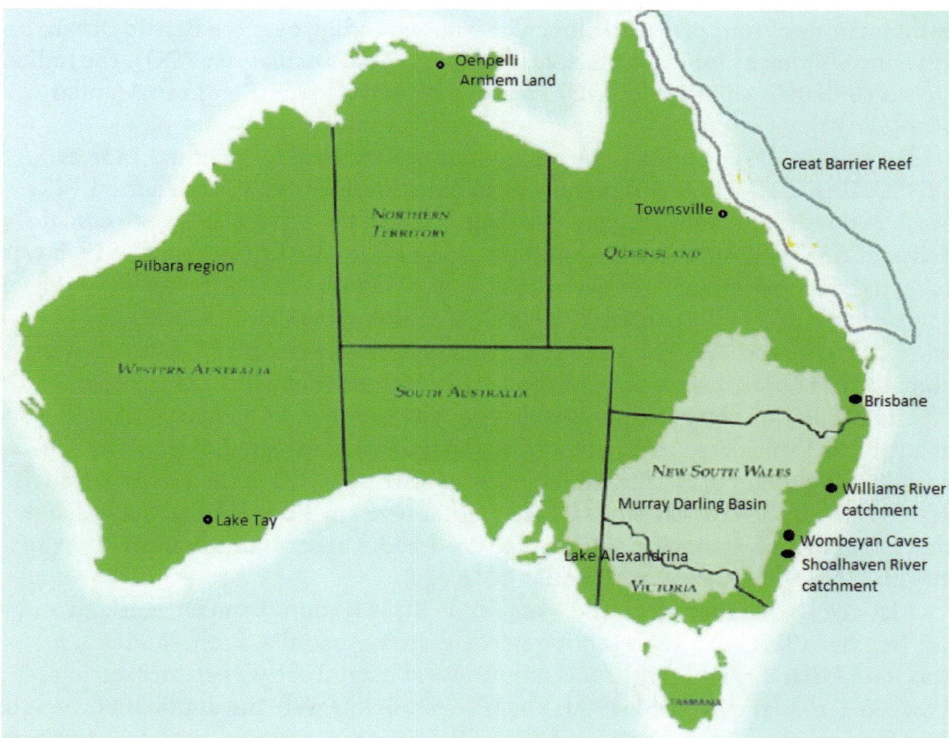

Figure 1.
Map of Australia showing states and locations referred to in this chapter.

2. Drought in Australia

A book by McKernan entitled *Drought: The Red Marauder* [5] reveals a story as perceived by people who have experienced droughts in Australia throughout more than 200 years of European settlement. At any particular time, there is often a drought somewhere in continent of Australia. However, there is often a reluctance to acknowledge drought as a persistent aspect of Australian life and the arrival of drought is often greeted with surprise [5] with a tendency for each drought to be perceived as "the worst on record" [5]. Droughts are a recurrent and natural part of the Australian climate, with evidence of drought dating back thousands of years. However, deficiencies in our capability to adequately monitor, attribute, forecast and manage drought are exposed whenever a drought occurs [6].

2.1 Drought frequency and relationship to climate indices

For most Australian regions, individual climate drivers, associated with particular climate indices, generally account for less than 20% of monthly rainfall variability [7]. It is, therefore, unlikely that a single climate phenomenon is responsible for all drought events. It is probable that different periods of extended drought are driven by different and/or multiple combinations of climatic processes [8, 9].

Three periods of prolonged droughts have occurred in south-east Australia during the period of instrumental records. These are known as the 'Federation drought' (1895–1902), 'World War II drought' (1937–1945) and the 'Big Dry' (1997–2010). Verdon-Kidd and Kiem [10] showed that these major droughts were related to the combinations of four principal climate drivers extending over the Pacific Ocean (El Nino Southern Oscillation: ENSO; Pacific Decadal Oscillation: PDO), the Indian Ocean (Indian Ocean Dipole: IOD) and the Southern Ocean (Southern Annular Mode: SAM).

Hiepp et al. [11] examined the relative contributions of four climate indices (ENSO, IPO, DMI and SAM) on rainfall in New South Wales. For the period 1948–2006, the study investigated the magnitude of the influence of each climate driver and its interaction on the rainfall at 15 locations distributed over NSW. It was reported that the influence of each driver at a particular site is different although some generalised patterns were evident. The results show that the ENSO has wide influence across over the entire state of New South Wales and is the primary climate driver of influence at 10 of the 15 sites analysed. The IPO (Interdecadal Pacific Oscillation) by itself does not have evidence for an influence on rainfall at any of the considered sites, but is influential when considered in combination with ENSO. Palmer et al. showed the importance of the IPO in modulating drought across Australia over past centuries [12]. Other investigations [13] have suggested that about half of Australian interannual-to-decadal precipitation variability may originate from as far away as the Atlantic Ocean.

McGree [14] examined rainfall data for 36 BOM stations from Queensland and the Northern Territory, representing north-eastern Australia. From this data, it was found that drought frequency, duration and magnitude was greater during 1981–2010 than during 1951–1980. The IPO and ENSO were the dominant drivers of drought occurrence over the period 1951–2010. The increase was not linear and was in a large part due to low-frequency variability, namely the positive phase of the IPO from 1977 to 1998. The switch to the negative phase of the IPO from 1999 resulted in a decade from 2000 with reduced drought activity. Carvalho et al. examined rainfall variability in the Shoalhaven river catchment in southern NSW and its relation to climatic indices [15]. This study found that although drought in this region is related to El Niño years, there was only a weak positive correlation between catchment rainfall and SOI, which was moderately enhanced during negative phases of the IPO.

Kiem and Franks investigated multi-decadal variability of drought risk by examining the performance of the Grahamstown Reservoir in NSW [16], calculating the probability of the reservoir storage level falling below 30% during three different phases of the IPO. This probability was almost 20 times greater during the positive IPO phase than it was when negative.

2.2 Drought and climate change

Although some reports [17] claim links between climate change and recent droughts in Australia with a high degree of certainty, an examination of the scientific literature would suggest a more cautious approach to stating unambiguous conclusions, due to the lack of rigorous evidence. Studies by Cai et al. [18] found that although climate models generally suggest that Australia's Millennium Drought was mostly due to natural multi-decadal variability, some late-twentieth-century changes in climate that influence regional rainfall are partially attributable to anthropogenic greenhouse warming. Cook et al. [19] examined the Millennium Drought during 2003–2009 and the record-breaking rainfall and flooding in austral summer 2010–2011 in eastern Australia. They found limited evidence for a climate change contribution to these events, but such analyses are restricted by the lack of information on long-term natural variability. Analysing a reconstruction of summer (December–January–February), they reported moisture deficits during the Millennium Drought fall within the range of the last 500 years of natural variability. van Dijik et al. [20] studied the Millennium Drought in southeast Australia (2001–2009). They found that prevailing El Nino conditions explained about two-thirds of rainfall deficit in east Australia, but the results for south Australia were inconclusive with a contribution from global climate change plausible, but unproven.

2.3 Impacts of drought

The impacts of droughts have been categorised as meteorological, hydrological, agricultural and socioeconomic [21]. Many of the reported studies have emphasised the impact of drought on agriculture, and this is a topic of current focus with drought in eastern Australia affecting many farming communities. Sheng and Xu [22] estimated that Millennium drought between 2002 and 2010 reduced agricultural total factor productivity by about 18% in Australia over the period.

Studies have been directed towards evaluating the impact of climate change on agricultural production in Australia. For example, drought frequently limits Australian wheat production, and the expected future increase in temperatures and rainfall variability will further challenge the productivity [23]. Relationships between wheat yields and climatic factors including rainfall are known to be complex and the subject of ongoing investigations [24]. Studies by Hunt et al. [25] show that the reduced yields of wheat associated with lower rainfall can be offset with adaptation through early sowing of the crop.

Feng [26] examined the impacts of rainfall extremes on wheat yield in semi-arid cropping systems in eastern Australia and found that the frequent shortages of rainfall in eastern Australia created a greater threat to crop growth than excessive rainfall.

There are many impacts of droughts other than agriculture [20]. For example, droughts in Australia have had effects on wildlife populations including waterbirds [27, 28]. Studies have shown that droughts have an effect on mental health of the population, particularly in rural areas of Australia [29]. Li et al. investigated the ecological effects of extreme drought [30], including water acidification and eutrophication in the Lower Lakes (Lakes Alexandrina and Albert) in South Australia.

3. Flooding rains

Flooding rains are also recurring feature of the Australian climate. For example, prolonged rainfall over large areas of Queensland led to flooding of historic proportions in December 2010, extending into January 2011 [31, 32]. About 33 people died as a result of those floods, with more than 78% of the state (an area larger than France and Germany combined) declared a disaster zone. More than 2.5 million people were affected [31] with approximately 29,000 homes and businesses experiencing some form of inundation, with the cost of flooding estimated to be over A$5 billion [31]. In January 2011, Brisbane, the state capital of Queensland, experienced its second highest flood in over a century. Major flooding occurred throughout most of the Brisbane River catchment, with an estimated 18,000 properties inundated [32]. More recently in 2018, extreme rainfall conditions inundated the city of Townsville, located in coastal north Queensland, experiencing flooding of large parts of the urban area [3, 33].

In addition to the impacts on urban infrastructure, floods may have a substantial impact on the ecosystem. For example, runoff following extreme rainfall has been associated with detrimental impacts on coral of the Great Barrier Reef [34]. Flooding can also impact on the establishment of tree seedlings [35] and has been implicated in the dieback of mangroves in Queensland rivers, rather than the effect of herbicides as suggested previously [36].

With the devastating impacts of floods, there is interest in understanding and potentially improving predictive capabilities. Studies by McMahon suggest that floods in south-eastern Queensland do not occur randomly but are associated with a repeating 40 year cycle [37].

4. Predicting droughts and floods

In addition to general circulation models (GCMs) [38–40] that attempt to implement physical models of climatic systems, considerable research has been reported over the past decade using machine learning, particularly neural networks, to forecast rainfall [41–48]. The results suggest that the skill of the forecasts using the machine learning approach for medium-term forecast is superior to GCMs [48]. With GCMs, forecasts are usually initially generated for extended grid areas for defined geographical areas. Forecasts for more specific locations can then be generated through a process of downscaling. With the machine learning approach, forecasts are generated for specific locations for which historical data are available. If there are sufficient locations over a geographical region, contour maps representing observed and forecast rainfall can then be generated. This is illustrated for Tasmania in **Figures 2** and **3** for observed and forecast monthly rainfall, 12 months, in advance.

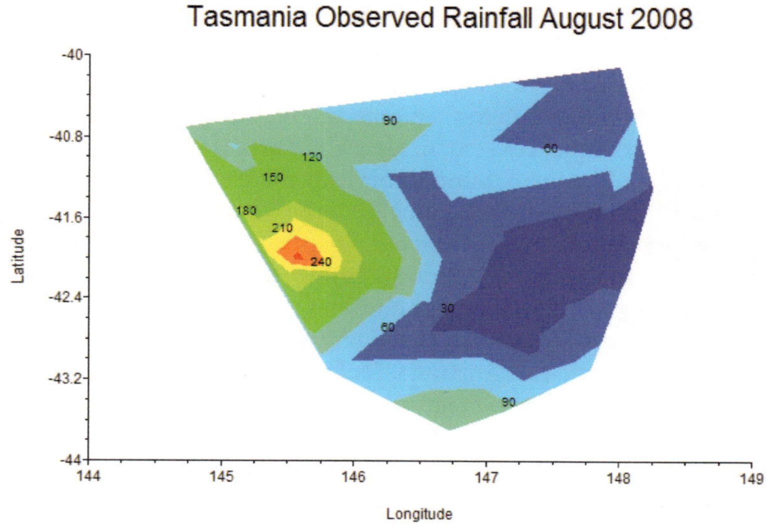

Figure 2.
Observed monthly rainfall for Tasmania August 2008.

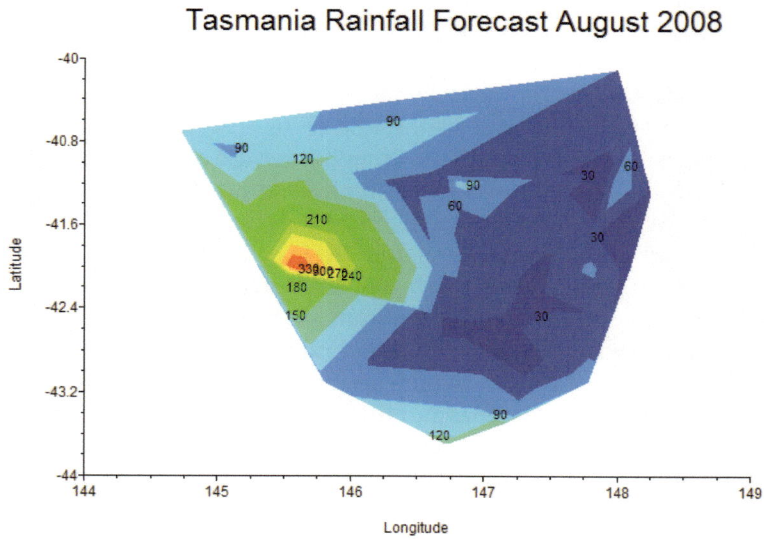

Figure 3.
Forecast monthly rainfall for Tasmania August 2008.

5. Rainfall reconstructions

Instrumental records of rainfall and temperature generally extend back only about 100 years in Australia. Reconstructions of past temperatures, extending back hundreds or thousands of years, are available for many parts of the world [49]. These are derived from palaeo data, including tree rings, corals, ice cores and stalagmites. Such reconstructions are much rarer for rainfall and, comparatively, few exist currently for Australia. However, it is important to consider the examples that exist as they enable episodes of droughts and flooding rains to be put into a wider context before asserting particular events are unprecedented, or ascribing them, with a high degree of certainty, to anthropogenic climate change.

Climate proxy data are ideally derived from sources that are located within, or in close proximity to, the region of interest. However, in cases where such proxy records are unavailable, remote proxies can be considered as a possible alternative [50]. There are relatively few in situ rainfall-sensitive palaeoclimate proxy records in Australia providing continuous records of rainfall variability [51]. An alternative is to utilise records that are from the same continent, but external to the region of direct interest. For example, Ho et al. made use of three such records [52]. The first was a study by Lough [53], who found significant correlations between coral luminescence intensity recorded in coral cores from the Great Barrier Reef and summer rainfall variability in northeast Queensland, enabling the multi-century coral record to be used to reconstruct Queensland summer rainfall back to the eighteenth century. The second study used high-resolution (approximately annual) analysis of trace elements sensitive to moisture availability present in a stalagmite from the Wombeyan Caves in south-east Australia. McDonald [54] found this to correlate with periods of above or below average rainfall from 749 BCE (before the Common Era) to 2001 CE (Common Era). The third study generated a 350 year long rainfall reconstruction at Lake Tay in Western Australia, based on tree ring widths [55].

These three Australian palaeoclimate proxies of rainfall enabled the reconstructions of rainfall in the Murray Darling Basin of south-eastern Australia [49, 50] although all three lie outside the Murray Darling Basin. The results reveal several extended periods that are likely to have been drier than indicated by the instrumental record from approximately the last century. Extended dry periods include the mid-late-1700s, 1500s, 1100s, 400s and 300 BCE. Comparisons between the reconstructed rainfalls and extreme instrumental rainfall indicated that the occurrence of extended periods wetter than the wettest decade in the instrumental record is also likely to have occurred in the mid-late-1800s and also around 1700. Multi-centennial wet epochs (or, at least, epochs without a multi-year drought) are also evident between 400–700 CE and 300 BCE–2100 CE. The reconstructions, therefore, confirm the occurrence during the last 2751 years of both wet and dry periods that have greater frequency, duration and severity than observations from the instrumental record.

O'Donnell [56] developed a 210-year tree ring-width chronology from *Callitris columellaris* from the Pilbara region of Western Australia. This was highly correlated with summer-autumn (December–May) rainfall across semi-arid northwest Australia. The reconstruction showed the periods of below average precipitation extending from one to three decades and the periods of above average precipitation which were often less than a decade. The results demonstrate that recent decades (1995–2012) have been unusually wet with average summer/autumn rainfall of 310 mm compared with the previous two centuries (average summer/autumn rainfall of 229 mm).

Freunmd et al. [57] used a diverse set of Southern Hemisphere palaeoclimate records to produce rainfall reconstructions for cool (April–September) and warm (October–March) seasons corresponding to eight regions across the Australian continent. They reported that trends towards wetter conditions in tropical northern Australia are highly unusual in the context of multi-century rainfall reconstructions. Cool season drying trends during the instrumental period in regions of southern Australia are very unusual, although not unprecedented, when compared with the past several centuries from 1600 CE.

Verdon-Kidd et al. [58] produced a 507 year reconstruction of rainfall for the monsoonal northwest of Australia, focussing on the site of Oenpelli in the Northern Territory. The study used remote proxies from Asia (tree rings), Australia (coral) and South America (tree rings) with an instrumental calibration period from 1900 to 1976. The rainfall reconstruction presented from 1470 CE suggests that the modern instrumental record on average represents a wetter climate than the pre-instrumental period. Furthermore, the reconstructions display wet and dry periods of greater duration than evident from the instrumental record. Other recent studies using tree rings to reconstruct rainfall include Allen [59] for Arnhem Land in the Northern Territory of monsoonal Australia, and O'Donnell for Western Australia [60]. Evidence is also provided from sediments from northwest Australia periods of extreme flooding and drought over past 2000 years [61].

A reconstruction of rainfall was produced by Tozer [62] for the Williams River catchment in subtropical eastern Australia (see map) extending over a period of 1013 years between 1000 and 2012 CE. Ho [50] relied on circulation teleconnections that strongly link climatic processes in one region to another. The remote proxies are calibrated with respect to an instrumental period, to develop palaeoclimate reconstructions. As no high-resolution palaeoclimate proxies were available corresponding to the Williams River catchment area, the study utilised the teleconnection between summer sea salt deposition recorded in ice cores from Law Dome in East Antarctica and rainfall variability in eastern Australia. It was found that both the dry and wet epochs persisted up to twice as long in the pre-instrumental compared with the instrumental period.

Documentary evidence can also be valuable in complementing palaeo reconstructions of rainfall. Fenby et al. [63, 64] considered 12 documentary-based rainfall chronologies for five subregions of south-eastern Australia (SEA) over the 1788–1860 period using a range of historical sources. This analysis identified 27 drought years in south-east Australia between 1788 and 1860 and 14 years of high rainfall in New South Wales (NSW) between 1788 and 1840. This study confirms that south-east Australia has experienced considerable rainfall variability that has influenced past Australian societies since the first European settlement in 1788. Of the droughts identified in this study, 1837–1841 was the longest and most widespread event influencing all subregions. The 1793–1809 period was particularly wet, with periods of heavy rainfall often resulting in devastating floods on the Hawkesbury River region of NSW.

6. Conclusion

This chapter provides some background to the available data and understanding of rainfall patterns experienced in Australia. Compared with many counties, Australia is fortunate in having a comparatively good set of rainfall records over the continent extending back about 100 years in many cases. The continent experiences extended episodes of drought and flooding rains. It is clear that these are related to multiple climate drivers that may extend widely over the globe, and are incompletely understood. Forecasting rainfall over the medium-to-long term remains problematical, and solutions may be found in advanced data analysis techniques such as machine learning rather than physical models. There remains a scarcity of rainfall reconstructions based on palaeo evidence enabling records to be extended back multiple centuries beyond the instrumental data. These are very important as they enable more recent episodes of drought and flooding rains to be placed into context.

Without this, there is a tendency in the general community to believe that a particular event is unprecedented because something similar did not occur in recent decades. The reconstructions are also important from the perspective of assigning rainfall patterns to climate change and designating as natural or anthropogenic in origin.

Acknowledgements

This work was supported by the B. Macie Family Foundation.

Author details

John Abbot[1,2,3]

1 Institute of Public Affairs, Melbourne, Victoria, Australia

2 University of Tasmania, Tasmania, Australia

3 Climate Modelling Laboratory, Noosa, Queensland, Australia

*Address all correspondence to: johnwabbot@gmail.com

References

[1] 119 Years of Australian Rainfall. Australian Bureau of Meteorology. Available from: http://www.bom.gov.au/climate/history/rainfall/

[2] Revealing South-Eastern Australia's Rainfall History. Available from: http://climatehistory.com.au/wp-content/uploads/2009/12/Revealing-Australias-rainfall.pdf

[3] Available from: https://www.abc.net.au/news/2019-02-15/queensland-floods-special-climate-statement/10816184

[4] State of the Drought Shows Dams Empty and NSW Drowning in Dust. Available from: https://www.abc.net.au/news/2019-03-12/state-of-the-drought-is-not-good/10876716

[5] McKernan M. Drought: The Red Marauder. Crows Nest, NSW, Australia: Allen and Unwin; 2005. p. 312

[6] Kiem AS, Johnson F, Westra S, et al. Natural hazards in Australia: Droughts. Climatic Change. 2016;**139**(1):37-54

[7] Risbey JS, Pook MJ, McIntosh PC, et al. On the remote drivers of rainfall variability in Australia. Monthly Weather Review. 2009;**137**:3233-3253

[8] Nicholls N. Local and remote causes of the southern Australian autumn-winter rainfall decline, 1958-2007. Climate Dynamics. 2010;**34**(6):835-845

[9] Kiem AS, Verdon-Kidd DC. Towards understanding hydroclimatic change in Victoria, Australia—Why was the last decade so dry? Hydrology and Earth System Sciences Discussions. 2009;**6**:6181-6206

[10] Verdon-Kidd D, Kiem A. Nature and causes of protracted droughts in Southeast Australia: Comparison between the federation, WWII, and big dry droughts. Geophysical Research Letters. 2009;**36**(22):L22707

[11] Hiep ND, Rivett K, MacSween K, et al. Association of climate drivers with rainfall in New South Wales, Australia, using Bayesian model averaging. Theoretical and Applied Climatology. 2017;**127**(1-2):169-185

[12] Palmer JG, Cook ER, Turney CSM, et al. Drought variability in the eastern Australia and New Zealand summer drought atlas (ANZDA, CE 1500-2012) modulated by the interdecadal Pacific oscillation. Environmental Research Letters. 2015;**10**(12):124002

[13] Johnson ZF, Chikamoto Y, Luo JJ, et al. Ocean impacts on Australian interannual to decadal precipitation variability. Climate. 2018;**6**(3):61

[14] McGree S, Schreider S, Kuleshov Y. Trends and variability in droughts in the Pacific Islands and Northeast Australia. Journal of Climate. 2016;**29**(23):8377-8397

[15] Carvalho RC, Woodroffe CD. Rainfall variability in the Shoalhaven River catchment and its relation to climatic indices. Water Resources Management. 2015;**29**(14):4963-4976

[16] Kiem AS, Franks SW. Multi-decadal variability of drought risk, eastern Australia. Hydrological Processes. 2004;**18**(11):2039-2050

[17] Steffen W, Rice M, Hughes L, et al. Climate Change and Drought. Australia: Climate Council of Australia Factsheet. June 2018. Available from: https://www.climatecouncil.org.au/resources/climate-change-and-drought-factsheet/

[18] Cai W, Purich A, Cowan T, et al. Did climate change-induced rainfall trends contribute to the Australian millennium drought? Journal of Climate. 2014;**27**(9):3145-3168

[19] Cook BI, Palmer JG, Cook ER, et al. The paleoclimate context and

future trajectory of extreme summer hydroclimate in eastern Australia. Journal of Geophysical Research-Atmospheres. 2016;**121**(21):12820-12838

[20] van Dijk AIJ, Beck HE, Crosbie RS, et al. The millennium drought in Southeast Australia (2001-2009): Natural and human causes and implications for water resources, ecosystems, economy, and society. Water Resources Research. 2013;**49**(2):1040-1057

[21] Mishra AK, Singh VP. A review of drought concepts. Journal of Hydrology. 2010;**391**(1-2):202-216

[22] Sheng Y, Xu X. The productivity impact of climate change: Evidence from Australia's millennium drought. Economic Modelling. 2019;**76**:182-191

[23] Watson J, Zheng B, Chapman S, et al. Projected impact of future climate on water-stress patterns across the Australian wheatbelt. Journal of Experimental Botany. 2017;**68**(21-22):5907-5921

[24] Yates D, Vervoort RW, Minasny B, et al. The history of using rainfall data to improve production in the grain industry in Australia from Goyder to ENSO. Crop & Pasture Science. 2016;**67**(5):467-479

[25] Hunt JR, Lilley JM, Trevaskis B, et al. Early sowing systems can boost Australian wheat yields despite recent climate change. Nature Climate Change. 2019;**9**(3):244

[26] Feng P, Wang B, Liu DL, et al. Impacts of rainfall extremes on wheat yield in semi-arid cropping systems in eastern Australia. Climatic Change. 2018;**147**(3-4):555-569

[27] Wen L, Saintilan N, Reid JRW, et al. Changes in distribution of waterbirds following prolonged drought reflect habitat availability in coastal and inland regions. Ecology and Evolution. 2016;**6**(18):6672-6689

[28] Johnston GR. Drought increases the impact of introduced European foxes on breeding Australian pelicans. Wildlife Research. 2016;**43**:507-514

[29] Hanigan IC, Schirmer J, Niyonsenga T. Drought and distress in south-eastern Australia. EcoHealth. 2018;**15**(3):642-655

[30] Li S, Bush RT, Mao R, et al. Extreme drought causes distinct water acidification and eutrophication in the lower lakes (Lakes Alexandrina and Albert), Australia. Journal of Hydrology. 2017;**544**:133-146

[31] Queensland Flood Commission of Enquiry, Final Report. 2012. Available from: http://www.floodcommission.qld.gov.au/__data/assets/pdf_file/0007/11698/QFCI-Final-Report-March-2012.pdf

[32] van den Honert RC, McAneney J. The 2011 Brisbane floods: Causes, impacts and implications. Watermark. 2011;**3**:1149-1173

[33] Reinisch S. Devastating impact on livestock and wildlife caught in the Townsville floods. Australian Veterinary Journal. 2019;**97**(4):N2

[34] Butler IR, Sommer B, Zann M, et al. The cumulative impacts of repeated heavy rainfall, flooding and altered water quality on the high-latitude coral reefs of Hervey Bay, Queensland, Australia. Marine Pollution Bulletin. 2015;**96**(1-2):356-367

[35] Maxwell A, Capon SJ, James CS. Effects of flooding on seedling establishment in two Australian riparian trees with contrasting distributions; Acacia stenophylla A. Cunn. ex Benth. and Casuarina cunninghamiana Miq. Ecohydrology. 2016;**9**(6):942-949

[36] Abbot J, Marohasy J. Has the herbicide diuron caused mangrove dieback? A re-examination of the evidence. Human and Ecological Risk Assessment. 2011;**17**(5):1077-1094

[37] McMahon GM, Kiem AS. Large floods in south East Queensland, Australia: Is it valid to assume they occur randomly? Australasian Journal of Water Resources. 2018;**22**(1):4-14

[38] Rocheta E, Sugiyanto ML, Johnson F, et al. How well do general circulation models represent low frequency rainfall variability? Water Resources Research. 2014;**50**(3):2108-2123

[39] Schepen A, Wang QJ, Robertson DE. Seasonal forecasts of Australian rainfall through calibration and bridging of coupled GCM outputs. Monthly Weather Review. 2014;**142**(5):1758-1770

[40] Hawthorne S, Wang QJ, Schepen A, et al. Effective use of general circulation model outputs for forecasting monthly rainfalls to long lead times. Water Resources Research. 2013;**49**:5427-5436

[41] Deo RC, Sahin M. Application of the artificial neural network model for prediction of monthly standardized precipitation and evapotranspiration index using hydrometeorological parameters and climate indices in eastern Australia. Atmospheric Research. 2015;**161**:65-81

[42] Feng P, Wang B, Liu DL, et al. Machine learning-based integration of remotely-sensed drought factors can improve the estimation of agricultural drought in south-eastern Australia. Agricultural Systems. 2019;**173**:303-316

[43] Ali M, Deo RC, Downs NJ, et al. Multi-stage committee based extreme learning machine model incorporating the influence of climate parameters and seasonality on drought forecasting. Computers and Electronics in Agriculture. 2018;**152**:149-165

[44] Deo RC, Sahin M. Application of the extreme learning machine algorithm for the prediction of monthly effective drought index in easternn Australia. Atmospheric Research. 2015;**153**:512-525

[45] Abbot J, Marohasy J. Application of artificial neural networks to rainfall forecasting in Queensland, Australia. Advances in Atmospheric Sciences. 2012;**29**(4):717-730

[46] Abbot J, Marohasy J. The potential benefits of using artiicial intelligence for monthly rainfall forecasting for the Bowen Basin, Queensland, Australia. Water resources management VII. WIT Transactions on Ecology and the Environment. 2013;**171**:287-297

[47] Abbot J, Marohasy J. Input selection and optimisation for monthly rainfall forecasting in Queensland, Australia, using artificial neural networks. Atmospheric Research. 2014;**138**:166-178

[48] Abbot J, Marohasy J. Chapter 3, engineering and mathematical topics in rainfall. In: Forecasting of Medium-Term Rainfall Using Artificial Neural Networks: Case Studies from Eastern Australia. Rijeka, Croatia: IntechOpen; 2018

[49] Abbot J, Marohasy J. The application of machine learning for evaluating anthropogenic versus natural climate change. GeoResJ. 2017;**14**:36-46

[50] Ho M, Verdon-Kidd DC, Kiem AS, et al. Broadening the spatial applicability of paleoclimate information-a case study for the Murray-Darling basin, Australia. Journal of Climate. 2014;**27**(7):2477-2495

[51] Ho M, Kiem AS, Verdon-Kidd DC. A paleoclimate rainfall reconstruction in the Murray-Darling basin (MDB), Australia: 2. Assessing hydroclimatic risk using paleoclimate records of wet and dry epochs. Water Resources Research. 2015;**51**(10):8380-8396

[52] Ho M, Kiem AS, Verdon-Kidd DC. A paleoclimate rainfall reconstruction in the Murray-Darling basin (MDB), Australia: 1. Evaluation of different paleoclimate archives, rainfall networks, and reconstruction techniques. Water Resources Research. 2015;**51**(10):8362-8379

[53] Lough JM. Great barrier reef coral luminescence reveals rainfall variability over North-Eastern Australia since the 17th century. Paleoceanography. 2011;**26**:PA2201

[54] McDonald JR, Drysdale Q, Hua E, et al. A 1,500 Year Southeast Australian Rainfall Record Based on Speleothem Hydrological Proxies, Paper Presented at AMOS Annual Conference 2013—Sense and Sensitivity: Understanding our Changing Weather and Climate, Melbourne, Australia.

[55] Cullen LE, Grierson PF. Multi-decadal scale variability in autumn-winter rainfall in South-Western Australia since 1655 AD as reconstructed from tree rings of *Callitris columellaris*. Climate Dynamics. 2009;**33**(2-3):433-444

[56] O'Donnell AJ, Cook ER, Palmer JG, et al. Tree rings show recent high summer-autumn precipitation in Northwest Australia is unprecedented within the last two centuries. PLoS One. 2015;**10**(6):e0128533

[57] Freund M, Henley BJ, Karoly DJ, et al. Multi-century cool- and warm-season rainfall reconstructions for Australia's major climatic regions. Climate of the Past. 2017;**13**:1751-1770

[58] Verdon-Kidd DC, Hancock GR, Lowry JB. A. 507-year rainfall and runoff reconstruction for the monsoonal north west, Australia derived from remote paleoclimate archives. Global and Planetary Change. 2017;**158**:21-35

[59] Allen KJ, Brookhouse M, French BJ. Two climate-sensitive tree-ring chronologies from Arnhem Land, monsoonal Australia. Austral Ecology. 2019;**44**:581-596

[60] O'Donnell AJ, Cook ER, Palmer JG, et al. Potential for tree rings to reveal spatial patterns of past drought variability across western Australia. Environmental Research Letters. 2018;**13**(2):024020

[61] Rouillard A, Skrzypek G, Turney C, et al. Evidence for extreme floods in arid subtropical Northwest Australia during the little ice age chronozone (CE 1400-1850). Quaternary Science Reviews. 2016;**144**:107-122

[62] Tozer CR, Vance TR, Roberts JL, et al. An ice core derived 1013-year catchment-scale annual rainfall reconstruction in subtropical eastern Australia. Hydrology and Earth System Sciences. 2016;**20**(5):1703-1717

[63] Fenby C, Gergis J. Rainfall variations in South-Eastern Australia part 1: Consolidating evidence from pre-instrumental documentary sources, 1788-1860. International Journal of Climatology. 2013;**33**(14):2956-2972

[64] Gergis J, Ashcroft L. Rainfall variations in South-Eastern Australia, part 2: A comparison of documentary, early instrumental and palaeoclimate records, 1788-2008. International Journal of Climatology. 2013;**33**(14):2973-2987

Section 2

Rainfall Patterns and Impacts

Chapter 2

Statistical Analysis of Rainfall Patterns in Jeddah City, KSA: Future Impacts

Mazen M. Abu Abdullah, Ahmed M. Youssef, Fawzy Nashar and Emad Abu AlFadail

Abstract

Recently, the Kingdom of Saudi Arabia (KSA) has been facing significant changes in rainstorm patterns (rainstorm intensities, frequencies, distributions) causing many flash flood events. The city of Jeddah is located in a coastal plain area, in the middle of the western side of the KSA, which represents a clear case of changing rainstorm patterns. Jeddah has been hit by many rainstorm events, which increased dramatically since 2009 (e.g., one in 2009, one in 2011, one in 2015, and another one happened in 2017). However, in 2018 about six rainstorms occurred. Two major flash flood events occurred in the city in November 2009 and in January 2011. There were significant impacts of these two events causing severe flooding. During these events, 113 persons were announced dead (in the 2009 event), and infrastructures and properties were damaged (roads and highways, more than 10,000 homes and 17,000 vehicles). In addition to that, dam failure occurred in the 2011 event. This situation gives clear evidence in changing the climate system that could cause more storms in the future across the KSA. Generally, Jeddah city has a lack of short-duration data in rainfall stations. In addition to that, there are a limited number of studies that have been done in determining rainstorm patterns. Consequently, the approach of the current study will focus on understanding and determining rainstorm patterns in the period between 2011 and 2017 depending on some digital rainfall stations that have been installed recently in Jeddah city. Rainstorm pattern and the method of distribution are the most crucial factors affecting peak flow and volume calculations. Our findings showed that there are two pattern types for the rainstorms in Jeddah city. Finally, a comparison with SCS-type II distribution was carried out.

Keywords: rainfall patterns, floods, impacts, Jeddah, KSA, statistical analysis

1. Introduction

Climate change is a debatable subject these days. Dealing with this topic is considered an enormous challenge of the coming years [1]. There are many definitions of climate; however, the common term of climate is the long-term pattern of meteorological conditions in a specific area [2]. It is measured by evaluating variations in temperature, humidity, atmospheric pressure, wind, precipitation, atmospheric particle, and other meteorological variables. Climate change can have significant impacts on weather conditions around the world, such as storms and heavy rainfall.

Climate changes can occur due to different processes internal to the earth, external forces (e.g., variations in sunlight intensity), and human activities that have been increasing recently. Many shreds of strong evidence in many regions regarding the climate changes and variability that is impacted by anthropogenic activities, industries and natural specifications of climate systems are recently available. Among these factors that caused climate change due to changing of the composition of the atmosphere are the greenhouse gas emissions, CO_2, CH_4, and N_2O [3–5].

Many studies related to climate change have emphasized that there is an increase in the rainfall events recently regarding frequency and intensity [6–14]. Rainfall patterns and rate over a region are reliant on the ambient and global water evaporation and to a significant extent on altitude, latitude, and level of humidity [15]. Warmer conditions brought out from increased emissions of greenhouse gases through industrialization cause evaporation and precipitation with varying degree of intensity on individual regions [16].

Various environmental impacts have been witnessed according to climate change including change in the hydrological cycles, availability of water resources, unprecedented rainfalls and floods, unexpected drought frequencies, and changes in natural ecosystems [5, 14, 17–20]. Rain is an essential natural phenomenon which can influence the human life and properties. There are many factors which affect rainfall, such as geographical position, monsoon, topographic, and other factors. Flash flood frequency and severity in the desert areas are generally unpredictable and vary from year to year due to variability in the rainfall values [21–23]. Rainfall in arid areas is spatially variable than that of humid regions and is often described as "spotty," and the impacted area is often limited by the radius of the clouds [24].

It is tough for hydrologists to use unreliable hydrological data (rainfall data) in the design of water-related structures. In most cases, the available rainfall data are limited (few records) and contain some gaps in the time series; rainfall stations are far from each other, with no intensity records; and records are not authentic values due to human errors. Previous studies used historical information to carry out the rainfall frequency analysis to understand the flooding behavior [25, 26]. These studies generally show that the use of historical information can be of great value in the reduction of the uncertainty in flood quantity estimators. A frequency analysis of the data is the most commonly applied method. Several studies were found dealing with the analysis of rainfall intensity in many areas [27–32].

A rain gauge is an instrument that measures how much rain falls in a given set time. Automatic rain gauges are rain gauges that electronically start working once it feels rain on the gauge. They automatically record the data, from measuring to removing the rainfall afterward [33]. Automatic rain gauge systems are required to collect rainfall data at remote locations, especially oceanic sites where logistics prevent regular visits [34]. It is usually based on tipping-bucket rainfall sensor and data logger for measuring the rainfall quantity and intensity during a given period and transmits the data through the GSM/GPRS modem to the desired e-mails and server at user-defined time intervals, and records obtained are of high reliability.

The Kingdom of Saudi Arabia (KSA) is one of the other countries that is impacted by severe events of rainfall in the last decade due to climate change. The intensity and frequency of the rainfalls are unprecedented and cause devastating floods in many KSA regions. Due to the lack of short-duration data in the Kingdom in general and in Jeddah in particular, and the limited number of studies that have determined the pattern of rainstorms, the current research will be a cornerstone in establishing rainfall pattern and behaviors. Determination of rainstorm pattern and method of distribution is one of the most critical factors affecting peak flow calculations. In this work, the Jeddah area was chosen as a unique example of unprecedented rainfall events in the last decade. The rainfall rate and patterns of the Jeddah area will be discussed in detail to extract the adequate rainfall intensity patterns that could be used for future predictions. In the current study, the short duration rainfall data recorded by climate stations of the Saudi Geological Survey during the period from 2011 to 2017, as well as the data available at station J134 of the Ministry of Environment, Water and Agriculture, And the conclusion of the general pattern prevailing in these storms, as well as their comparison with the distribution of SCS-Type II, and the conclusion of the intensity curves based on available data. Many rainfall storms were analyzed in this study. It includes the following events (intensity records) 2011, 2014, 2015, 2016 and 2017. However, in the devastating floods that occur in 2009, we do not have intensity records for that event.

2. Rainfall distribution in the KSA and recent problems

The KSA climate is mainly arid and semiarid, except in the northern and southern areas. The arid and semiarid regions have an extreme continental climate with warm and dry summer and very cold winter especially in the central regions of the KSA. To assess rainfall pattern over the study area, it is necessary to define the dominant climatic patterns that have an influence on rainfall distribution over the western province of the KSA. The climatic pattern can best be described by considering the various air masses that affect the rainfall distribution over the KSA area. The influence of the different air masses and the rainfall patterns over the KSA has been discussed and mapped by several investigators [35–38].

Different air masses, which influence the Kingdom's climate, are illustrated in **Figure 1**. These air masses include (1) the monsoon front during the late autumn (maritime tropical air mass) reaches the area from the south, southwest, and southeast. This front that originates in the Indian Ocean and the Arabian Sea during the autumn brings warm and moist air. Outbreaks of westerly air become more frequent, characterized by medium to high intensity over the western and northwesterly regions of the country. This front often picks up further moisture while moving through the Red Sea Trough. (2) The continental tropical air masses are warm and moist coming from the Atlantic Ocean through the Central and North African continent. (3) The maritime polar air masses are derived from the Eastern Mediterranean

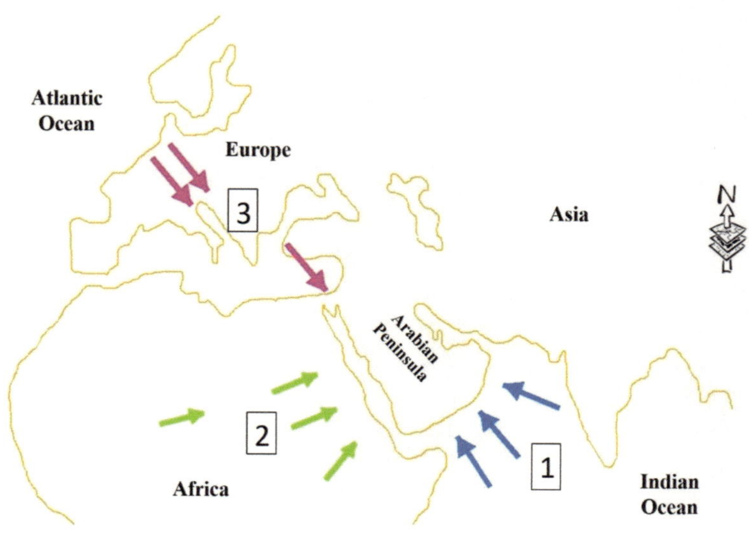

Figure 1.
Air masses affecting the climate of the KSA: (1) maritime tropical air masses, (2) continental tropical air masses, and (3) maritime polar air masses [39].

Sea. In early winter the Mediterranean-borne maritime air increasingly disturbs the monsoonal air movement and displaces it in the low altitudes. These maritime depressions draw the tropical continental air masses into the warm sectors, and extreme weather conditions occur that are associated with the passage of a hot sector. Both (2) and (3) move toward the east and prevail in the winter season. During this season, the western region, particularly the coastal area, is characterized by its relatively low rate of rainfall [38], whereas, due to the topographic effects, the highlands receive a considerable amount of rainfall. In spring, the impact of the Mediterranean air movement diminishes, whereas the monsoon from the south takes its place, penetrating the southern part of the Kingdom. During summer, the cyclonic flow sweeps along the Mediterranean Sea from the west toward the east and continues moving over the northern and central regions of the country preventing the maritime air masses of the northeasterly monsoon from penetrating the north regions of the Kingdom. Due to this, the summer season will be somewhat dry in the area considered.

The mean annual rainfall (from 1960 to 2018) over the KSA can be shown in **Figure 2**. The rainfall rate in the KSA (except for the Empty Quarter desert) is 109 mm per year. The southwestern region is characterized by a heavy rainfall compared with the rest of the Kingdom. Rainfall is more than 500 mm per year in some locations. The peaks of the Sarawat Mountains, which extend from the northern part of the Kingdom to the south, parallel to the Red Sea, are the dividing line for the distribution of surface water. The rains that fall in the east of this line take place in the valleys heading east, and in the west, the valleys are steep and flowing towards the Red Sea.

Recently, there is a definite increasing trend in rainfall in frequency and intensity in the KSA. Many areas in the KSA have shown an increasing trend in the annual rainfall and flood events. Most of the rainfall in the KSA occurs during the monsoon. Monsoon is a term derived from the Arabic word "mausim," meaning season. It was used to describe the seasonal winds of the Arabian Sea. The Jeddah area is characterized by wet and dry seasons that generally occur from November to May and from June to October, respectively. One example is Jeddah city which has shown an increasing trend in the rainfall events. These events cause disasters that result in human, property, and economic loss.

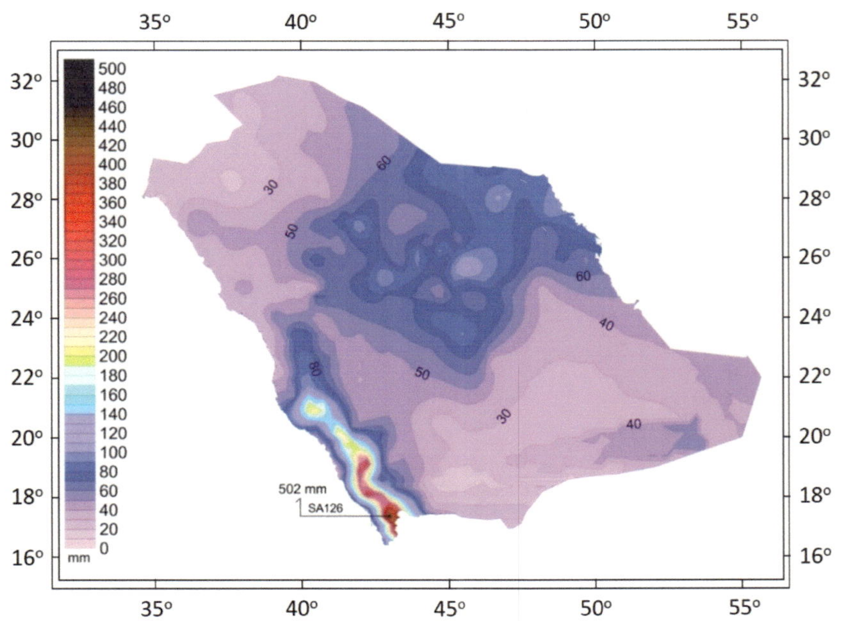

Figure 2.
Average annual rainfall values for 270 rainfall stations distributed across the KSA area (from 1960 to 2018).

These unprecedented events have affected the KSA, causing considerable damage to highways, railroads, urban zones, and agricultural areas [39]. Most of the flash flood hazards in the KSA are caused by a combination of natural conditions (heavy rainfall and climate changes) and human interference (poor drainage systems and urban expansion). Recently, heavy rainfall events have triggered flash floods in various areas of the KSA (e.g., Jeddah city flash floods in the years 2009, 2011, 2015, 2017, and 2018 and Al Riyadh flash floods in 2015 and 2018). The severely hit areas are generally in the western part of the KSA, particularly in the city of Jeddah during November 2009 and January 2011 [39, 40]. These events were characterized by 70 and 111 mm of rainfall, respectively, within 3 h and were considered catastrophic flash floods for Jeddah city. They caused a death toll of 113 people in 2009, and, together, they damaged more than 10,000 homes and destroyed approximately 17,000 vehicles. Other areas impacted by flash floods are As Sail Al Kabir area, Taef; Najran city; Ar Riyadh city; Ha'il city; Makkah city; and Tabuk city [41–48].

3. Study area and its characteristics

The study area, the Jeddah area, is located in the western region of the KSA covering an area of ~1731 km^2 and lies between latitudes 21°15′ and 21°57′N and

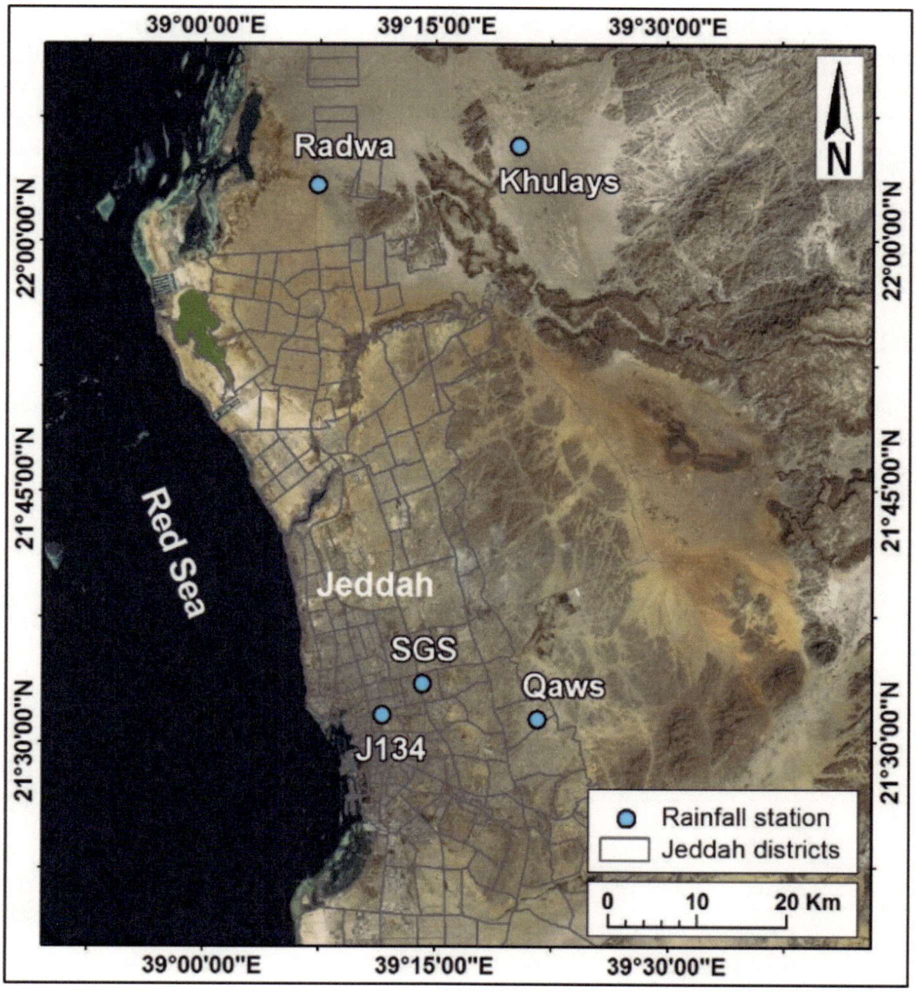

Figure 3.
Study area and its meteorology station locations.

longitudes 39°06′ and 39°31′E (**Figure 3**). From the latitude point of view, the Jeddah area is considered to be in the arid zone (Köppen's climate classification). The Jeddah area has different geomorphological features. It represents a part of the Red Sea coastal zone. Jeddah drainage system comes from the east (a series of hills) toward the city which is located to the west. The catchment areas have a variety of landforms such as low- and midsize hills and flattened foothills in some places followed to the west by the floodplain areas. The monthly average relative humidity ranges between ~85% from September to October and ~34% from April to June. The Jeddah area is characterized by scarce rainfalls with high variability, which occur most often in November–December–January (the winter season). Additionally, monsoonal rainstorms take place in March and April due to the moist air currents from the Indian Ocean and the Arabian Sea. The average annual precipitation is ~52.5 mm/year. The maximum rainfall was recorded in 1996, with ~284 mm/year. Recently these thunderstorms increased in their frequencies and intensities causing flash floods and inundation events within Jeddah city.

4. Characteristics

In this study, all the current stations have automatic rain gauges that can record the intensity of the rainfall events (minute-base). This technique did not occur before, and most of the hydrological studies were depending on the daily rainfall data analysis which brings some uncertainty to the hydrological results. The stations used in the current study are shown in **Figure 3**. In the present work, trend and pattern analysis is based on intensity rainfall values for each rainfall station. Detailed analysis and discussion of the different recently occurred storms will be carried out in the following sections. This analysis and comparison work will give us the ability to deduce the general pattern prevailing in the storms in the city of Jeddah. **Table 1** has a list of different storms that will be considered in this study.

4.1 Rainstorm occurred on January 26, 2011

The data recorded by the rainfall station located at the headquarter of the Saudi Geological Survey shows that the precipitation began at 10:56 am on Saturday, January 26, 2011; continued until 12:27 noon, for ~91 min; and remained after a short break until 3:38 pm. The total precipitation is ~112.0 mm. It is difficult

No.	Date	Location	Name	Total rainfall (mm)	Duration (h)
1	January 26, 2011	SGS site	SGS	112.0	4.7
2	November 22, 2014	SGS site	SGS	43.0	10.0
3	November 17, 2015	Radwa farms	Radwa	21.7	2.7
4	November 17, 2015	Khulays, Ad Daff	Khulays	20.6	2.7
5	November 17, 2015	Old Jeddah airport	J134	79.0	3.2
6	November 17, 2015	Wadi Qaws	Qaws	61.0	1.77
7	December 02, 2016	SGS site	SGS	44.91	2.75
8	November 21, 2017	SGS site	SGS	88.04	15.75

Table 1.
Rainstorms occurred between 2011 and 2017 in the Jeddah area and the stations' names.

to distinguish two separate storms, but this rainstorm can be separated into two phases. The first phase of rainfall was ~54 mm, while the second one was ~58 mm. **Figure 4a** shows the rainfall recorded plotted against the whole rainfall time (4.7 h), which shows ~48% of the rainfall depth is fallen in the first phase for a period of ~78 min. The first phase is shown in **Figure 4b**.

4.2 Rainstorm occurred on November 22, 2014

This event was recorded in one rainfall station located at the headquarter of the Saudi Geological Survey. The data recorded by this station shows that the precipitation began at 0:22 am on Saturday, November 22, 2014; continued until 1:03 am, for ~41 min; then resumed at 9:53; and stayed until 10:19 am. The total precipitation is ~42.87 mm. This rainstorm can be separated into two storms. The rainfall of the first storm was ~33 mm, while the second one was ~9.61 mm. **Figure 5a** shows the rainfall recorded plotted against the whole rainfall time (10 h), which shows ~77% of the rainfall depth is fallen in the first storm for a period of ~41 min. The first storm is shown in **Figure 5b**.

4.3 Rainstorm occurred on November 17, 2015

This event was recorded in four rainfall stations including Radwa farms, Wadi Khulays, Station J134, and Wadi Qaws.

The data recorded by the rainfall station at Radwa farms shows that the precipitation began with a slow rate at 8:44 am on Tuesday, November 17, 2015; continued until 9:51 am, for ~67 min; and remained with a moderate rate until 11:25 am of the same day. The total precipitation is ~21.7 mm. The main storm in this event lasted ~41 min and reached a precipitation amount of ~19 mm. **Figure 6a1** shows the rainfall recorded plotted against the whole rainfall time (2.7 h), which shows ~88% of the rainfall depth is fallen at a period of ~41 min (the main storm; see **Figure 6a2**).

The data recorded by the rainfall station at Wadi Khulays shows that the precipitation began at 9:18 am on Tuesday, November 17, 2015; continued until 9:45 am, for ~27 min; and continued intermittently until 12:00 noon. The total precipitation is ~20.6 mm. The main storm in this station lasted ~27 min and reached a precipitation amount of ~18 mm. **Figure 6b1** shows the rainfall recorded plotted against the whole rainfall time (2.7 h), which shows ~87% of the rainfall depth is fallen at a period of ~27 min. The first storm is shown in **Figure 6b2**.

The data recorded by rainfall station J134 (located in the old Jeddah airport) shows that the precipitation began at 10:00 am on Tuesday, November 17, 2015; continued until 12:20 am, for ~140 min; and continued intermittently until

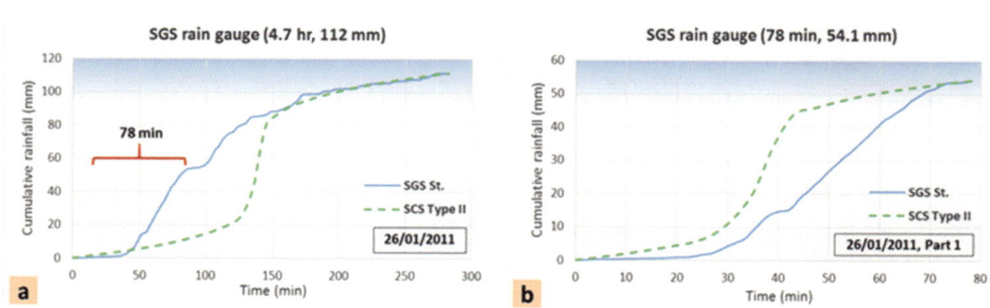

Figure 4.
The rainstorm on January 26, 2011, at SGS station. (a) Distribution of cumulative rainfall intensity of the total storm. (b) Distribution of cumulative rainfall intensity of the first phase.

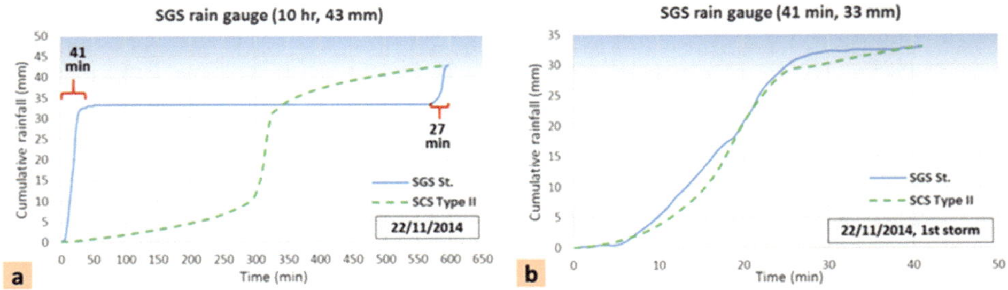

Figure 5.
The rainstorm on November 22, 2014, at SGS station. (a) Distribution of cumulative rainfall intensity of the total storm. (b) Distribution of cumulative rainfall intensity of the first storm.

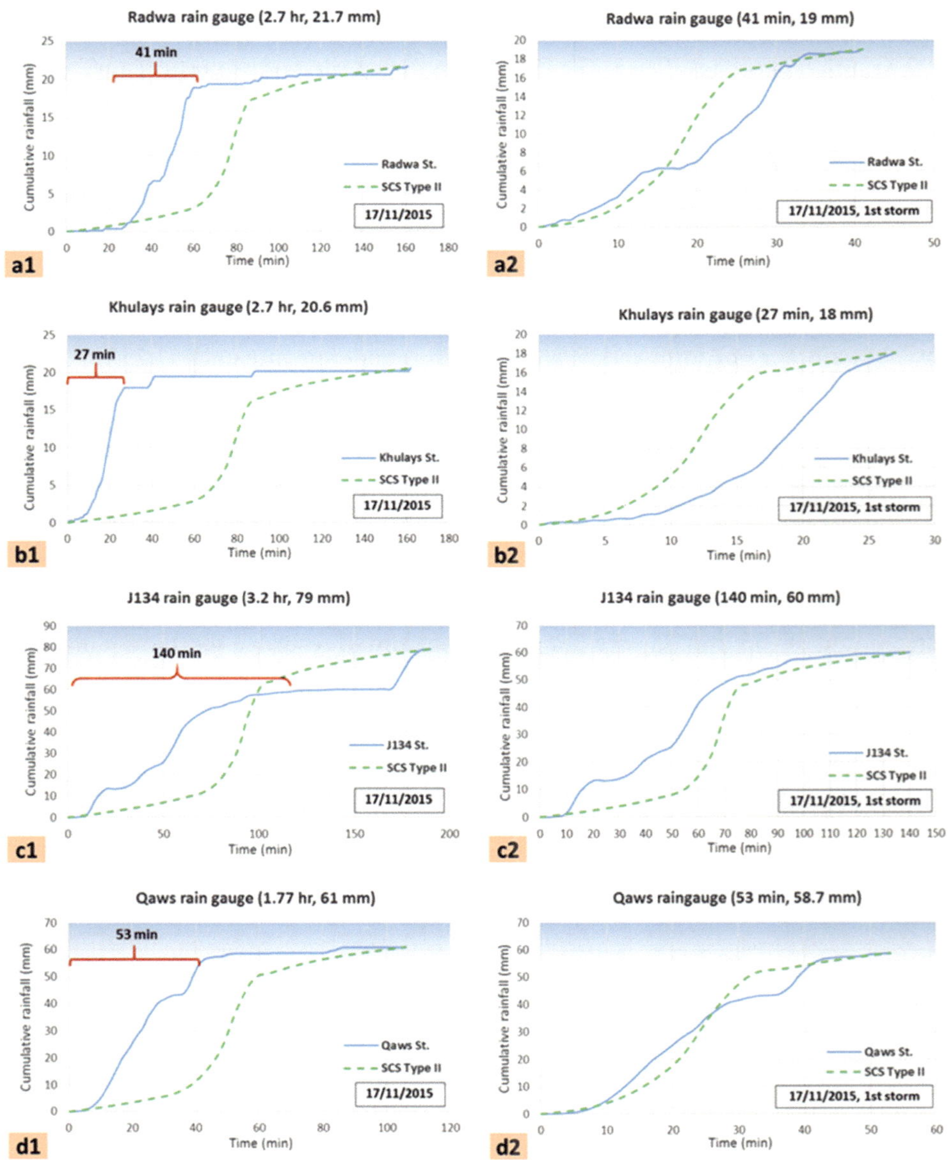

Figure 6.
The rainstorm on November 17, 2015, at Radwa, Khulays, J134, and Qaws stations. (a1, b1, c1, d1) Distribution of cumulative rainfall intensity of the total storm. (a2, b2, c2, d2) Distribution of cumulative rainfall intensity of the first storm.

1:10 pm. The total precipitation is ~79.0 mm. The main storm in this station lasted ~140 min and reached a precipitation amount of ~60 mm. **Figure 6c1** shows the rainfall recorded plotted against the whole rainfall time (3.2 h), which shows ~76% of the rainfall depth is fallen at a period of ~140 min. The first storm is shown in **Figure 6c2.**

The data recorded by the rainfall station at Wadi Qaws shows that the precipitation began at 10:48 am on Tuesday, November 17, 2015; continued until 11:41 am, for ~53 min; halted at 12:09; and then continued intermittently until 12:34 pm. The total precipitation is ~61.0 mm. The main storm in this station lasted ~53 min and reached a precipitation amount of ~58.7 mm. **Figure 6d1** shows the rainfall recorded plotted against the whole rainfall time (1.8 h), which shows ~96% of the rainfall depth is fallen at a period of ~53 min. The first storm is shown in **Figure 6d2.**

4.4 Rainstorm occurred on December 2, 2016

This event was recorded in one rainfall station located at the headquarters of the Saudi Geological Survey. The data recorded by this station shows that the precipitation began at 7:18 am on Friday, December 2, 2016 and continued until 7:38 am, for ~20 min; then it resumed at 8:11 and stayed until 8:43 am, followed by a precipitation of 0.25 mm at 10.03 am. The total rainfall is ~44.91 mm. This rainstorm can be separated into two storms. The rainfall of the first storm was ~12.43 mm, while the second one was ~31.47 mm. **Figure 7a** shows the rainfall recorded plotted against the whole rainfall time (2.75 h), which shows ~28% of the rainfall depth is fallen in the first storm for a period of ~20 min and the second storm which lasted 43 min shows 70% of the total rainfall depth. **Figure 7b** shows the main storm which lasted 43 min (the second storm).

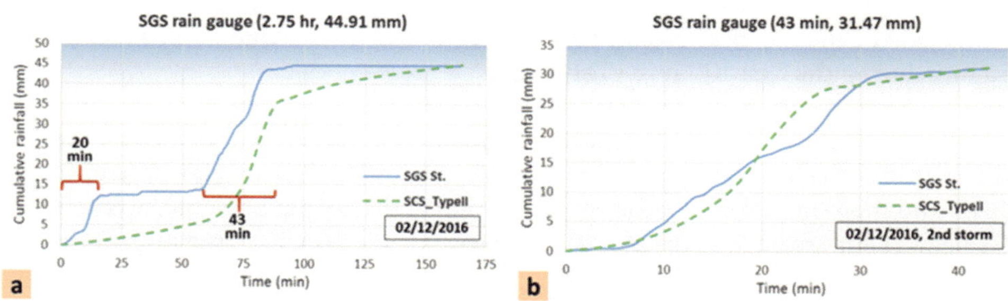

Figure 7.
The rainstorm on December 2, 2016 at SGS station. (a) Distribution of cumulative rainfall intensity of the total storm. (b) Distribution of cumulative rainfall intensity of the second storm.

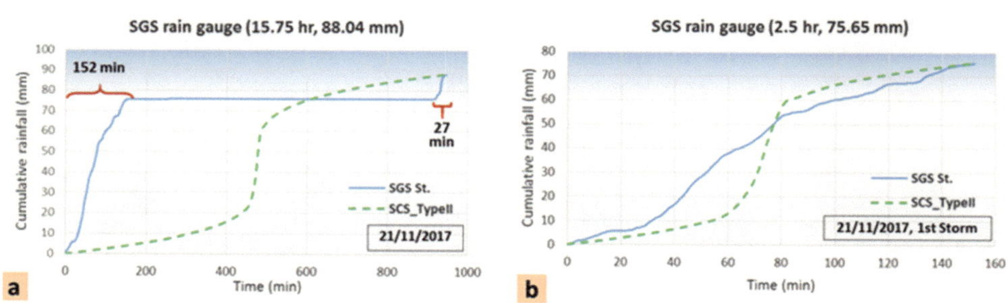

Figure 8.
The rainstorm on November 21, 2017, at SGS station. (a) Distribution of cumulative rainfall intensity of the total storm. (b) Distribution of cumulative rainfall intensity of the first storm.

4.5 Rainstorm occurred on November 21, 2017

This event was recorded in one rainfall station located at the headquarters of the Saudi Geological Survey. The data recorded by this station shows that the precipitation began at 8:19 am on Tuesday, November 21, 2017; continued until 10:50 am, for ~152 min; then resumed at 11:30 pm; and continued until 12:01 night. The total precipitation is ~88.04 mm. This rainstorm can be separated into two storms. The rainfall of the first storm was ~76 mm, while the second one was ~12.0 mm. **Figure 8a** shows the rainfall recorded plotted against the whole rainfall time (15.75 h), which shows ~85% of the rainfall depth is fallen in the first storm for a period of ~152 min. The first storm is shown in **Figure 8b**.

5. Comparison of different storms

In the current work, a comparison between the storms which occurred in the Jeddah area during the period (2011–2017) was carried out. The results indicated that there is some similarity in the behavior of these storms. The presence of the two main storms characterizes most of them during the rainfall time which is usually separated by a period of partial cessation. The most substantial amount of rainfall always exceeding 70% of the recorded rainfall appears in the first storm. Also, it was found that there is one exception in this rule which seems in the storm recorded on January 26, 2011, at the Saudi Geological Survey Station, which is different from other storms in terms of the amount of precipitation where the rain value reached 112 mm. **Figure 9a** shows a comparison of the major storms that have been occurred in the Jeddah area. **Figure 9b** shows comparison between storms which exceeded a rain value (50 mm). The results also indicated that based on the rainstorm that hit the city of Jeddah on November 21, 2017, with the storms that had a definite impact in the Jeddah city, we find that all these storms exceeded the amount of rainfall 50 mm during the first 100 min (**Table 2**).

The storm recorded at Wadi Qaws station on November 17, 2011, is the highest intensity, followed by the storms occurred at J134 on November 17, 2015; then the station at the headquarters of the Saudi Geological Survey on November 21, 2017; and then the storm recorded at the station at the headquarter of the Saudi Geological Survey on January 26, 2011. However, the storm occurred on January 26, 2011, left a massive damage and problems to the Jeddah area. It caused failure of the Umm al-Khair dam and paralyzed traffic in the province. Maybe the reason for that

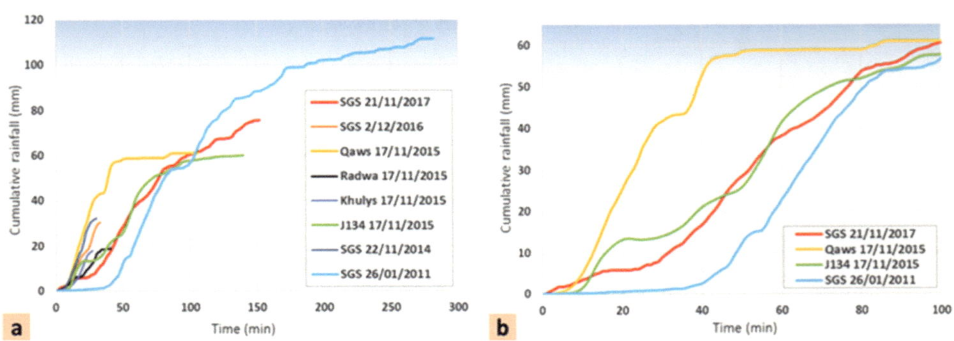

Figure 9.
A comparison between different rainstorms occurred in the period between 2011 and 2017: (a) all storms and (b) storms recorded rainfall more than 50 mm (the first 100 min).

is related to the duration time of the storm on January 26, 2011. The impact of the storm on November 21, 2017, was within the city of Jeddah and led to the flooding of a large number of roads and tunnels, although considers moderate intensity, however, the storm was concentrated in the center of the city, where almost urban areas are located. The lack of a proper drainage system inside the city has increased the impact of this storm.

6. Prevailing patterns of precipitation in Jeddah

To deduce the characteristic patterns of rainfall in the Jeddah area, we need a large number of reordering storms which are not available in the whole KSA. However, in the current work depending on the storms collected in the Jeddah area from the new installed automated rainfall stations, some suggestions can be made to understand the behavior of storms in the Jeddah area. To achieve this, the data available for each storm were converted to dimensionless data (**Figure 10**). The behavior of all possible storms was compared with the SCS-type II distribution curve. It is noticed that these storms can be divided into two groups. The first group represents the storms, which exceeded 50 mm of precipitation (**Figure 11a**). This group is represented by the storms recorded at the headquarters of the SGS station on November 21, 2017; January 26, 2011; and November 17, 2011, in both the Wadi Qaws and J134 stations. The second group represents short-term storms which recorded precipitation quantities less than 50 mm (**Figure 11b**). This group includes the storms registered at the headquarters of the SGS station on November 22, 2014, and December 2, 2016. For both groups, the average curves of each storm set were derived. To verify that these two storm groups are different from each other and have different behaviors, a distribution histogram of the average curve for each storm group was extracted and simplified as shown in **Figure 12**.

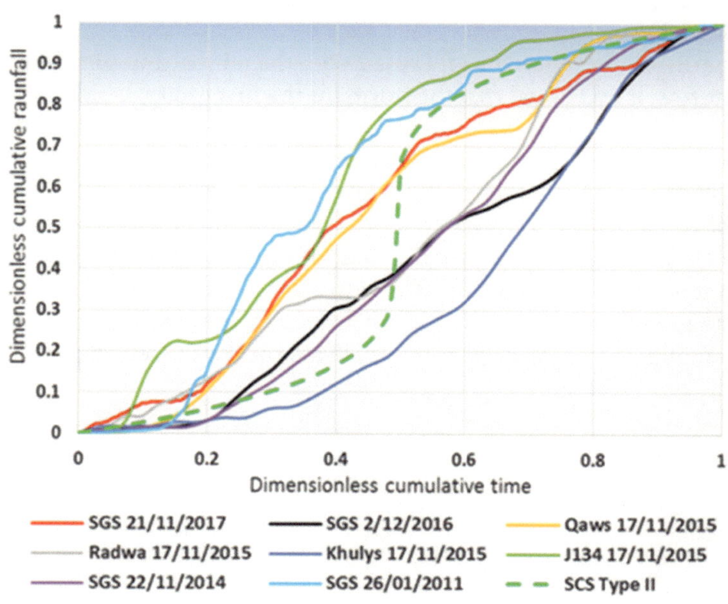

Figure 10.
Cumulative dimensionless hyetographs for rainstorms occurred in the period 2011–2017.

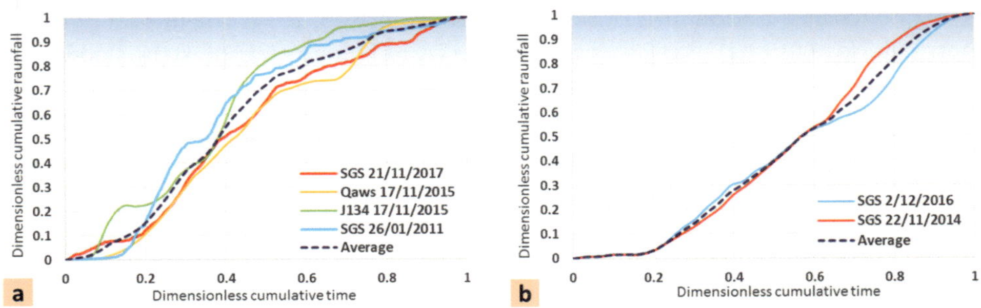

Figure 11.
Cumulative dimensionless hyetographs: (a) for storms above 50 mm and (b) for storms below 50 mm.

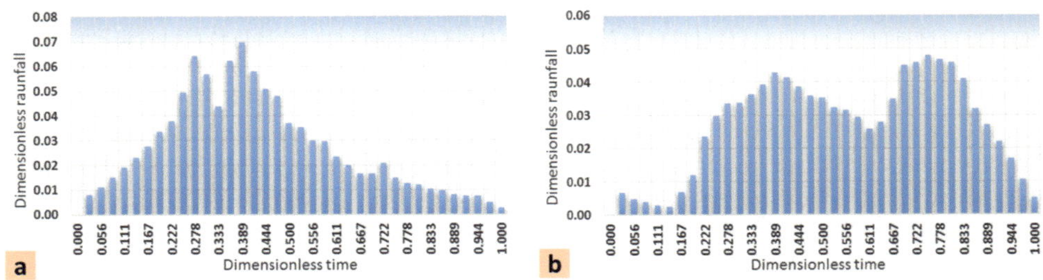

Figure 12.
(a) Dimensionless histogram for storms above 50 mm. (b) Dimensionless histogram for storms below 50 mm.

7. The relation of the storms to the depth of the rain calculated for different return periods in the Jeddah area

Rainfall depth of return periods 5–100 years was estimated by the Saudi Geological Survey and AECOM company (**Table 2**). The values are due to the analysis of the historical records of all rain stations located in and around the Jeddah area. An average value for each return period was calculated and used in the current study.

By comparing the average data of **Table 2** with the rainfall amounts recorded in the recent storms, it was found that the storms of the first group, which exceeded the amount of rainfall of 55 mm, fall within the rainfall depth above the 5-year return period. The storm recorded on January 26, 2011, 111.6 mm, falls below the 50-year return period; the storm recorded on November 21, 2017, 88 mm, falls below the rainfall value of the 20-year return period; the storm recorded on November 17, 2015, in the Wadi Qaws, 61 mm, falls under the value of the 10-year return period; and the storm recorded on November 17, 2011, at the J134 station, 79 mm, falls under the storms of the 12-year return period, while in the second group storms, they did not exceed the amount of rainfall of 50 mm, which falls below the return period of 5 years.

Source	**Return period**				
	5	**10**	**25**	**50**	**100**
SGS 2016 (mm)	53.1	71.1	93.4	110.0	125.0
AECOM 2011 (mm)	57.6	75.6	98.4	115.0	132.0
Average (mm)	**55.4**	**73.4**	**95.9**	**112.5**	**128.5**

Table 2.
Rainfall analysis of the historical records for the Jeddah area.

Accordingly, our results indicated that instead of using the SCS distribution curve type II which has happened before in our studies, we can use these new distribution curves. For storms that exceed 50 mm, we can use the distribution curve extracted from the first group (average curve) (**Figures 11a** and **12a**). However, for storms less than 50 mm, we can use the average distribution of the second group (small storm distribution can be applied) (**Figures 11b** and **12b**).

8. Conclusion

Our findings indicated that Jeddah rainfall is characterized by two patterns: one for the rain values less than 50 mm and the other one for the rain values above 50 mm. The study shows that dealing with rainfall data in the Jeddah area required the following points to be considered: (1) it is necessary to pay attention to the details of rainstorms (intensity values) and not only to the recorded daily values. (2) There is a convergence of the nature of storms with impact (greater than 30 mm), which begins to rise after the onset of the storm almost immediately and continue the same tendency to the end of the storm. (3) There is a slight difference between the nature of these storms and the SCS-type II distribution, which assumes that about 60% of the depth of the rain falls in a fraction of the duration of the storm ranging from 8 to 20% and the rest distributed over the rest of the storm. (4) The average length of storms should be considered within the study areas. (5) In short-term storms of less than 2 h or even (with some reservation) up to 3 h, the distribution of SCS type II can be used safely but with duration limited to the same span of the storm rather than 24 h.

Acknowledgements

The authors would like to thank the Saudi Geological Survey for providing scientific and logistical support for this work.

Conflict of interest

The authors declare no conflict of interest.

Author details

Mazen M. Abu Abdullah[1]*, Ahmed M. Youssef[1,2], Fawzy Nashar[3] and Emad Abu AlFadail[1]

1 Department of Geological Hazards, Applied Geology Sector, Saudi Geological Survey, Jeddah, KSA

2 Department of Geology, Faculty of Science, Sohag University, Egypt

3 Department of Hydrogeology, Applied Geology Sector, Saudi Geological Survey, Jeddah, KSA

*Address all correspondence to: mazenaa@yahoo.com

References

[1] Aerts J, Droogers P, editors. Climate Change in Contrasting River Basins: Adaptation Strategies for Water, Food, and Environment. London: CABI Books; 2004. p. 288

[2] Mander WJ. Dictionary of Global Climatic Change. 2nd ed. London: VCL; 1994

[3] Domroes M, El-Tantawi A. Recent temporal and spatial temperature changes in Egypt. International Journal of Climatology. 2005;**25**:51-63

[4] Hunt BG. Natural climatic variability and Sahelian rainfall trends. Global and Planetary Change. 2000;**24**:107-131

[5] Raziei T, Arasteh PD, Saghafian B. Annual rainfall trend in arid & semi-arid regions of Iran. In: ICID21st European Regional Conference; 15-19 May, 2005; Frankfurt (Oder) and Slubice—Germany and Poland; 2005. pp. 20-28

[6] Ali SA, Aadhar S, Shah HL, Mishra V. Projected increase in hydropower production in India under climate change. Scientific Reports. 2018;**8**:12450. DOI: 10.1038/s41598-018-30489-4

[7] Coumou D, Rahmstorf S. A decade of weather extremes. Nature Climate Change. 2012;**2**:491-496. DOI: 10.1038/nclimate1452

[8] Fowler HJ, Cooley D, Sain SR, Thurston M. Detecting change in UK extreme precipitation using results from the climate prediction.net BBC climate change experiment. Extremes. 2010;**13**:241-267. DOI: 10.1007/s10687-010-0101-y

[9] Kundzewicz ZW, Kanae S, Seneviratne SI, Handmer J, Nicholls N, Peduzzi P, et al. Flood risk and climate change: Global and regional perspectives. Hydrological Sciences Journal. 2014;**59**:1-28. DOI: 10.1080/02626667.2013.857411

[10] Mosmann V, Castro A, Fraile R, Dessens J, Sanchez JL. Detection of statistically significant trends in the summer precipitation of mainland Spain. Atmospheric Research. 2004;**70**:43-53

[11] Ogburn SP. Climate Change is Altering Rainfall Patterns Worldwide. Scientific American. 2013. Available from: http://www.scientificamerican.com/article/climate-change-isaltering-rainfall-patterns-worldwide/

[12] Xu ZX, Takeuchi K, Ishidaira H. Monotonic trend and step changes in Japanese precipitation. Journal of Hydrology. 2003;**279**:144-150

[13] Xu ZX, Takeuchi K, Ishidaira H, Li JY. Long-term trend analysis for precipitation in Asia Pacific friend river basin. Hydrological Processes. 2005;**19**:3517-3532

[14] Yu PS, Yang TC, Kuo CC. Evaluating long-term trends in annual and seasonal precipitation in Taiwan. Water Resources Management. 2006;**20**:1007-1023

[15] Nayak JK, Prajapati JA. Handbook on Energy Conscious Buildings, Pilot Edition. 2006. Available from: http://www.mnre.gov.in/solar-energy/ch2.pdf

[16] Stephen Schneider H. The greenhouse effect: Science and policy. Science. 1989;**243**:771-781

[17] González-Hidalgo JC, De Luis M, Raventos J, Sanchez JR. Spatial distribution of seasonal rainfall trends in a western Mediterranean area. International Journal of Climatology. 2001;**21**:843-860

[18] IPCC. Climate Change 2007: Synthesis Report. Spain: Valencia; 2007

[19] Jiang T, Su B, Hartmann H. Temporal spatial trends of precipitation & river flow in the Yangtze River Basin, 1961-2000. Geomorphology. 2006;**85**:143-154

[20] Piccareta M, Capolongo D, Boenzi F. Trend analysis of precipitation and drought in Basilicata from 1923 to 2000 within a southern Italy context. International Journal of Climatology. 2004;**24**:907-922

[21] Dolman AJ, Gash JHC, Goutorbe J-P, Kerr Y, Lebel T, Prince SD, et al. The role of the land surface in Sahelian climate: HAPEX-Sahel results and future research needs. Journal of Hydrology. Feb 1997;**188-189**:1067-1079

[22] Reid I, Powell DM, Laronne JB, Garcia C. Flash floods in desert rivers: Studying the unexpected. EOS. Transactions of the American Geophysical Union. 1994;**452**:75-39

[23] Warner TT. Desert Meteorology. Edinburgh: Cambridge University Press; 2004. p. 612

[24] Laity JE. Deserts and Desert Environments. Oxford, UK: Wiley-Blackwell; 2008. p. 360

[25] Ouarda TBMJ, Cunderlik JM, St-Hilaire A, Barbet M, Bruneau P, Bobee B. Data-based comparison of seasonality-based regional flood frequency methods. Journal of Hydrology. 2006;**330**:329-339

[26] Parent E, Bernier J. Bayesian POT modelling for historical data. Journal of Hydrology. 2003;**274**:95-108

[27] Dairaku K, Emori S, Oki T. Rainfall amount, intensity, duration and frequency relationships in the Mae Chaem watershed in Southeast Asia. Journal of Hydrometeorology. 2004;**5**:458-470

[28] De Paola F, Giugni M, Topa ME, Bucchignani E. Intensity-duration-frequency (IDF) rainfall curves, for data series and climate projection in African cities. Springerplus. 2014;**3**:133. DOI: 10.1186/2193-1801-3-133

[29] Lu E, Zhao W, Gong L, Chen H, Wang H, Li X, et al. Determining starting time and duration of extreme precipitation events based on intensity. Climate Research. 2015;**63**(1):31-41

[30] Mirhosseini G, Srivastava P, Stefanova L. The impact of climate change on rainfall intensity-duration-frequency (IDF) curves in Alabama. Regional Environmental Change. 2013;**13**:25-33. DOI: 10.1007/s10113-012-0375-5

[31] Wagesho N, Claire M. Analysis of rainfall intensity-duration-frequency relationship for Rwanda. Journal of Water Resource and Protection. 2016;**08**:706-723. DOI: 10.4236/jwarp.2016.87058

[32] Zainudini MA, Sardarzaei A. Analysis of short duration rainfall intensity data of Makoran region-Iran. Irrigation and Drainage Systems Engineering. 2014;**3**:123. DOI: 10.4172/2168-9768.1000123

[33] Goodison BE, Louie PYT, Yang D. World Meteorological Organization. WMO Solid Precipitation Measurement Intercomparison: Final Report. Instruments and Observing Methods Report No. 67 (WMO/TD-No. 872). Geneva: 1998

[34] Nystuen JA, Proni JR. A Comparison of Automatic Rain Gauges. Miami, Florida: NOAA/Atlantic Oceanographic and Meteorological Laboratory, Ocean Acoustics Division; 1995. DOI: 10.1175/1520-0426(1996)013<0062:ACOARG>2.0.CO;2

[35] Al-Ehaideb. Precipitation distribution in the southwest of Saudi Arabia [thesis]. USA: Arizona State University; 1985. p. 215

[36] Al-Qurashi MA. Synoptic climatology of the rainfall in the southwest region of Saudi Arabia [thesis]. USA: Western Michigan University; 1981. p. 97

[37] Alyamani MS, Sen Z. Regional variations of monthly rainfall amounts in the Kingdom of Saudi Arabia. Journal of King Abdulaziz University: Earth Science. 1992;**6**:113-133

[38] McLaren International Ltd. Water and agricultural development studies, Arabian Shield South. Report prepared for the Ministry of Water and Agriculture, Riyadh; 1979

[39] Sen Z. Hydrology of Saudi Arabia. In: Symposium on Water Resources in Saudi Arabia; Riyadh, Saudi Arabia: King Saud University; 1983. pp. 68-94

[40] Youssef AM, Pradhan B, Sefry SA. Flash flood susceptibility mapping in Jeddah city (Kingdom of Saudi Arabia) using bivariate and multivariate statistical models. Environment and Earth Science. 2016;**75**:12. DOI: 10.1007/s12665-015-4830-8

[41] Youssef AM, Sefry SA, Pradhan B, Abu Al Fadail E. Analysis on causes of flash flood in Jeddah city (Kingdom of Saudi Arabia) of 2009 and 2011 using multi-sensor remote sensing data and GIS. Geomatics, Natural Hazards and Risk. 2016;**7**:1018-1042. DOI: 10.1080/19475705.2015.1012750

[42] Abdul Karim AAA. Effect of spatial changes of urban growth and land uses on increasing flood risks in the Saudi City: Case study of Ha'il city using geographical information systems and remote sensing (GIS & RS). Arab Journal of Geographic Information Systems, Saudi Geographical Society. 2013;**6**(2)

[43] Abushandi E. Flash flood simulation for Tabuk City catchment, Saudi Arabia. Arabian Journal of Geosciences. 2016;**9**:188. DOI: 10.1007/s12517-015-2192-x

[44] Al-Ghamdi K, Elzahrany R, Mirza M, Dawod G. Impacts of urban growth on flood hazards in Makkah City, Saudi Arabia. International Journal of Water Resources and Environmental Engineering. 2012;**4**(2):23-34

[45] Elkhrachv I. Flash flood Hazard mapping using satellite images and GIS tools: A case study of Najran City, Kingdom of Saudi Arabia (KSA). The Egyptian Journal of Remote Sensing and Space Sciences. 2015;**18**:261-278. DOI: 10.1016/j.ejrs.2015.06.007

[46] El Shinnawy I, Bestawy A, El Tahawy T. Assessment and Management of flash floods for sustainable development in Al-Sail Al Kabir area, Kingdom of Saudi Arabia. International Journal of Applied Engineering Research. 2017;**12**:2807-2814

[47] Embaby A, Abu Halawa A, Ramadan M. Integrating geotechnical investigation with hydrological modeling for mitigation of expansive soil hazards in Tabuk City, Saudi Arabia. Open Journal of Modern Hydrology. 2017;**7**(1):11-37. DOI: 10.4236/ojmh.2017.71002

[48] Sharif HO, Al-Juaidi FH, Al-Othman A, Al-Dousary I, Fadda E, Jamal-Uddeen S, et al. Flood hazards in an urbanizing watershed in Riyadh, Saudi Arabia. Geomatics, Natural Hazards and Risk. 2014;**7**:702-720. DOI: 10.1080/19475705.2014.945101

Chapter 3

A Novel Assessment of the Impacts, Vulnerability, and Adaptation of Climate Change in Eastern Africa

Msafiri Yusuph Mkonda

Abstract

The evidences that climate change (CC) is a real situation have been established by different scholars and international organizations. However, much of the vast and burgeoning literature on CC has indicated spatial differences on the impacts, vulnerability, and adaptation among different communities. On that basis, various studies have grouped Eastern Africa as the most affected region by CC and has the weakest capacity to adapt or/and mitigate the dreadful conditions posed by CC. This chapter attempts to indicate the socioeconomic vulnerability in eastern Africa that has been coupled by climate impacts. In doing so, a wide range of studies reflecting the region has been reviewed. The results exhibit that there has been a cause-effect relationship between anthropogenic activities and climate impacts. Although both natural and anthropogenic factors cause climate change, the latter is more pronounced in local conditions. The level of deforestation and degradation in most eastern Africa is exceedingly high and this has subsequently increased the emission of the top green greenhouse gases (CO_2, CH_4, and N_2O) into the atmosphere. Thus, the resolutions of various conference of parties on CC need to adhere by both developed and developing countries for the betterment of the planet.

Keywords: agricultural production, adaptation policy, CC, eastern Africa, vulnerability

1. Introduction

Numerous international agreements on climate change, natural resources utilization, and development from the 2000s have made various calls to embolden scientific writing that would establish the concepts and good methodological approaches on climate vulnerability [1–3]. As a result, thousands of citable papers have been published in various reputable international journals (including *Environment, Development and Sustainability*) as responses to these calls [4]. However, due to increased vulnerability coupled with climate challenges in most developing countries, there have been increasing needs to establish more studies that give empirical evidences of the current climatic situation [4–6]. This aims to come up with tangible and scientific information relevant for discussion in various national and international dialogs to shape the future prospects. This is a basis for the Fifth Assessment Report (AR5) on the Intergovernmental Panel on Climate Change (IPCC) to convene a call for papers to elicit rigorous climatic finding in most developing countries especially the sub-Saharan Africa which is the most vulnerable region on Earth to climate change impacts [4].

Although the whole sub-Saharan Africa experiences unprecedented impacts of climate change, some regions face the worst compared to others [5–7]. Among the regions are Eastern Africa and the Horn of Africa [8, 9]. According to FAOSTAT [10], the Eastern African economy is highly dependent on agriculture, which is dominated by traditional rain-fed small-scale production, and thus, any further erratic weather tends to underpin crop failure in the region.

On the other hand, excessive droughts have ruthlessly affected animal husbandry in the region, and much of this effect happens in the already stressed areas like Central Tanzania, Northeastern Kenya, and the driest parts of the Horn of Africa. The vulnerability of the developing countries has been coupled by lack of strong institutions to deal with calamities and environmental disasters (**Table 1**). This has even increased the level of pollution and degradation as a search of alternative livelihoods. The Conference of the Parties (COP) on climate change, i.e., COP15 (Copenhagen in 2009), COP16 (Cancun in 2010), COP17 (Durban in 2011), COP18 (Doha in 2012), COP19 (Warsaw in 2013), and COP20 (Lima in 2014), realized this problem and, thus, proposes measures to reduce environment problem, more particularly the emission of greenhouse gases.

While various climate models from individual authors and the recent IPCC reports have confirmed that global climate change is real and that it is occurring more rapidly [4], there has been a need to establish empirical evidences that indicate the level of vulnerability and adaption especially in developing countries. This is particularly important as it is recognizable that over 66% of the global population is starving, whereas most of this population is in developing countries especially sub-Saharan Africa [3]. Thus, the understanding of actual climate scenarios in the region will inform the discussion for coping and mitigating the climate impacts.

As a response to the call of the Fifth Assessment Report of the IPCC regarding climate change [4], this paper attempts to discuss the causes, vulnerability, and adaptation and mitigation measures that exist in Eastern Africa in order to unveil the real climatic situation to various stakeholders at both local and global levels. This will enable climate practitioners to intensify the curbing of the top greenhouse gases such as carbon dioxide (CO_2), methane gas (CH_4), nitrous oxide (N_2O), and

Institutional factors	Economic factors	Environmental factors
i. Informal skills ii. Local knowledge iii. Formal education, skills, and technology iv. Informal network v. Formal security network vi. Strength of local institutions	i. Labor ii. Health iii. Access to natural resources iv. Access to communal resources v. Access to alternative economic opportunities	i. Risk environment ii. Degraded environment iii. High dependence of climate-sensitive sectors and natural resources iv. Communal lands and resources
Source: Modified from Eriksen and Noes [11].		

Table 1.
Examples of factors that influence vulnerability in the region.

chlorofluorocarbons (CFCs) through proper mitigation strategies for the sustainability of the planet [3]. Similarly, at the local level, the improved understanding among the farmers on the influence of climate on agricultural production is desirable for coping with actual and expected variations in both temperature and precipitation [12]. This will again curb an increasing number of undernourished people in the region.

Therefore, climate resilience at the local level is potentially optimized by proper adaptation measures coupled by appropriate and affordable mitigation measures. This underpins the effects posed by the temporal changing weather and climate in the region. Thus, proper adaptation measures that mainly focus on the agro-ecosystems are particularly useful to meet the demands of the increasing population in the region rather than using the "slash and burn" practices [13]. Eventually, the majority of the population lack livelihood option due to entitlement failure [14].

2. Location

Eastern Africa regions mostly cover Kenya, Uganda, Tanzania, Rwanda, Burundi, South Sudan, and other parts of the Horn of Africa that entail Eritrea, Ethiopia, Somalia, and Djibouti. This ecological region covers approximately 4,000,000 km^2, where Tanzania is the largest country and the smallest ones are Burundi and Djibouti. The region has a population of over 200 million. The region is bordered with two big water bodies, the Indian Ocean and the Red Sea. Tanzania, Kenya, Eritrea, Djibouti, and Somalia are bordered with the Indian Ocean and the Red Sea, while Ethiopia, Rwanda, Burundi, and Uganda are the landlocked

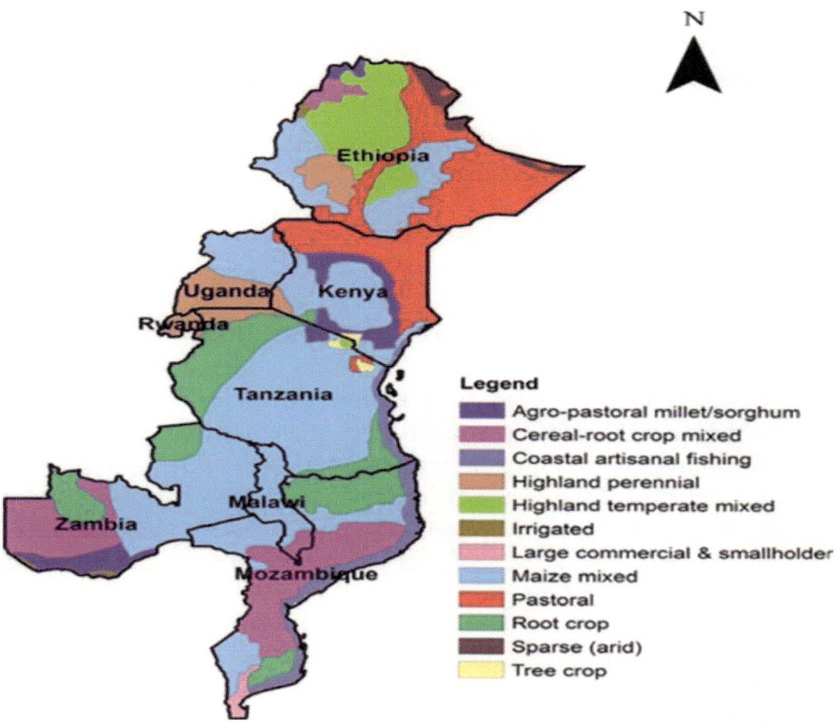

Figure 1.
Land use map of the study area. Source: [21].

countries in the region. Topographically, the Eastern African region is very complex, including mountains, valleys, rifts, rivers, ridges, and lakes.

The greatest rift in the world (Eastern Great Rift) is located in this region, and earthquake and a volcanic eruption are active [15]; the volcanic eruption of 2010 that occurred in the Danakil Depression of Ethiopia and Eritrea is a good example [16]. The elevation of the region ranges from the highlands of Ethiopia to the Danakil Depression below sea level; the Danakil Depression (local name Dallol) is the hottest place on the surface of the Earth (>60°C). Lakes Victoria, Tana, Abaya, and Turkana are the biggest surface water located along the rift valley [17–19]. The Lake Tana in Ethiopia is the source of Blue Nile River, which is a tributary of the Nile River [20]. Drought due to climate change and land gradation is the major environmental concern of this region (**Figure 1**).

3. Climate situation

Various studies from both local and international level indicate that climate change is a major agenda in the area [1]. This is important because Eastern Africa is among the worst vulnerable regions to climate change impacts [2, 3]. Although the region has been hit by this change, there are spatial differences on the magnitude of the impacts. The areas experiencing semiarid climates (especially Central Tanzania, Northeastern Kenya, and Uganda) suffer the most than those experiencing equatorial (Southern Uganda and Kenya and parts of Northern Tanzania) and tropical climates [12, 15, 22]. Overall, climate variability has been a critical driver of year-to-year impacts on both managed and unmanaged ecosystems in the area.

In order to properly understand its ecological impacts, it is essential to quantify how various ecosystems have historically responded to climate variability and to characterize the uncertainty in projected impacts. In this aspect, the general understanding dynamics of farmer adaptation and decision-making is particularly important. Now, the present study explores the major causes, vulnerability, impacts, and adaptations and mitigation measures. Rigorous literature has been significantly consulted to meet the objectives.

4. Major economic livelihoods

The major economic livelihood in the region is crop production, livestock keeping, and mixed crop-livestock production. In the latter agroecosystem, the crop residues are useful in supporting the feeding of livestock in addition to meadows and pastures. Likewise, animal manure is useful in fertilizing the farms. FAOSTAT [10] shows that crop production accounts for about 70%, while livestock accounts for about 30% of the total worth of agricultural production in the region. Despite being dominant among the pastoral communities only, the latter plays a key role both as a source of animal products and as a key input to production in the region. Among the dominant pastoral societies are the Maasai (Tanzania and Kenya), Turkana, Kalenjin, Kisii, Embu (Kenya), and Karamojong and Banyankole (Uganda) who mainly herd cattle, sheep, goats, and donkey for various purposes. Agricultural practices in the region are traditional, dominated by small-scale farms mainly under 2 ha per each household, and are characterized by low inputs of physical capital, fertilizers, and pesticides [23]. FAOSTAT [10] further observed that more than 90% of the agricultural production is rain-fed, thus making the practice not a dependable one on the face of the changing climate [3, 5, 12]. The self-sufficient ration is under 70% in most areas [24], therefore needing more food support from external sources.

5. Causes of climate

Despite the monotony of explaining the causes of climate change, it is fairly pretty to highlight these causes especially those with anthropogenic characteristics. Apart from natural causes that seem to have natural balance, anthropogenic activities in both developed and developing countries have been observed by various studies to have increased greenhouse gases in the atmosphere [3, 4]. In developing countries, deforestation and other forms of environmental degradation release thousands of tons of carbon in the atmosphere [18]. Subsequently, animal husbandry has been observed to emit lots of CH4 in the atmosphere. In addition, since most developing countries are dumpsite of various fabricated industrial goods, e.g., refrigerators which in turns emit CFCs, it is realizable that this system contributes significantly to the emission of this dangerous greenhouse gas. While that happens in developing countries, their counterparts (i.e., developed countries) emit more greenhouse gases through the pollution from industries. Since the focus of the present study is on developing countries, much of the examples and discussion will be recapped from the study area.

One can ask how the developing countries have significant contribution to greenhouse gases. The answer can be difficult especially on the quantification; however, the ways of doing so are obvious. The dominant agricultural systems and other livelihoods reflect the level of greenhouse emission from the case study area [3]. Since agriculture and animal husbandry are the major socioeconomic livelihoods in the areas, it is understandable that the expansion of agricultural farm and intensification of umber herds have significant contribution to greenhouse emission.

On the aspect of whether climate is changing or not, even most lobbyists and pessimists to climate change agree on the changes; however, they pose a zest for inquiry on whether the change is significant or needs to be ranked higher than other challenging factors.

6. Vulnerability of eastern Africa region to climate change

According to Adger [25], vulnerability is the state of susceptibility to harm from exposure to stresses associated with environmental and social change and from the absence of capacity to adapt. There are theories and conditions that subject a person or community into the state of vulnerability. Among the conditions that best explain this situation is entitlement failure [14]. Thus, the poor or weak people develop more vulnerability than rich people. Despite of being understandable, the challenges for vulnerability research are to develop robust and credible measures, to incorporate diverse methods that include perceptions of risk and vulnerability, and to incorporate governance research on the mechanisms that mediate vulnerability and promote adaptive action and resilience [25, 26].

We explore the state of vulnerability in the Eastern African region in order to propose synergies between vulnerability and on resilience of social ecological systems. This review will not only show the real contexts of the area but also optimize the adoption of suitable mitigation measures that would serve the purpose of both regional and international level. It provides evidence-based investigation to inform the discussion in the international arena and more particularly in the Conference of

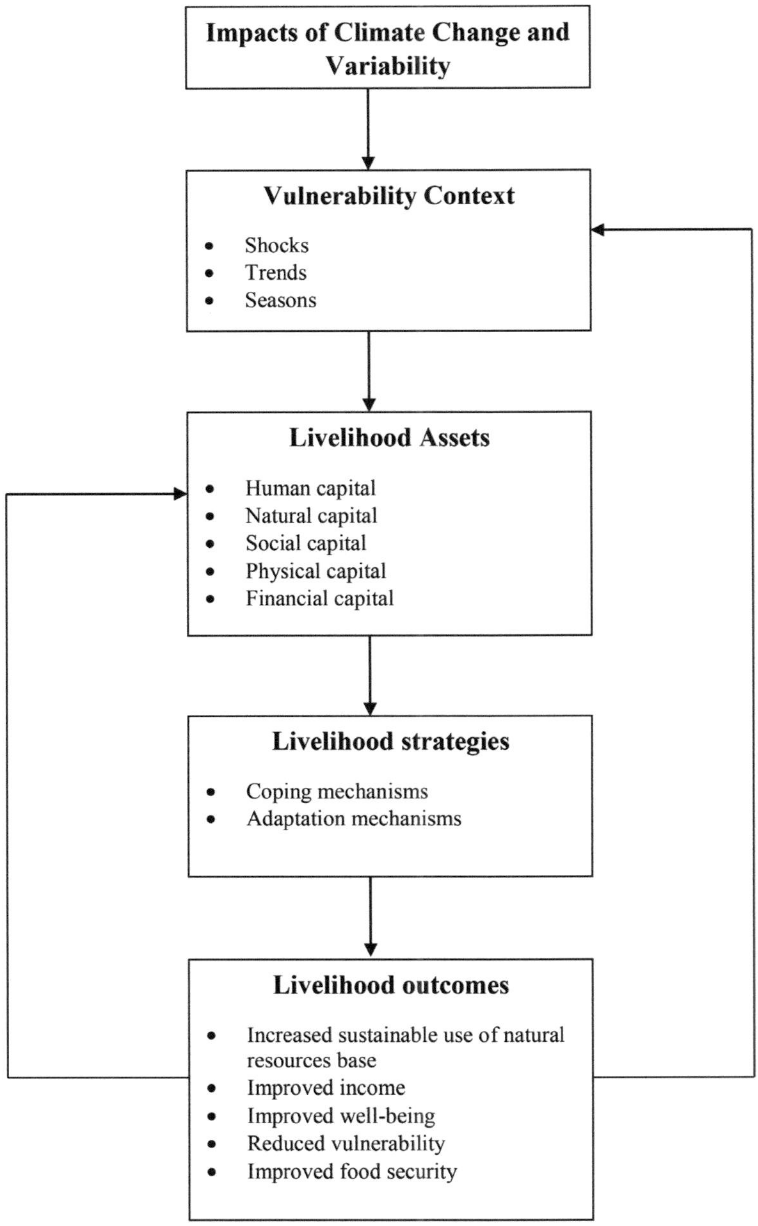

Figure 2.
Climate variability, vulnerability and adaptation, and its livelihood outcome. Source: Modified from the Sustainable Livelihood Framework [27].

the Parties on climate change and the subsequent reports by the Intergovernmental Panel on Climate Change [1–3]. **Figure 2** explores the overall concept of vulnerability to reduce it for best outcomes [27].

However, the measurement of vulnerability can be complicated by the fact that it emerges through complex interactions between biophysical and social dimensions across multiple scales, all of which vary across time, location, the nature of biophysical stressors, and outcomes of interest [28]. Therefore a combined theoretical and practical approach should be adopted in doing the same.

Further conceptualization reveals that in most developing countries, there have been factors that influence vulnerability for decades. These factors need to be controlled in order to increase the resilience of the farmers, thus increasing their capability in the production process.

Crane et al. [28] coined that for the vulnerability to be well assessed, there is a need to assess the systems of vulnerability, how this system is vulnerable, what are the causes of vulnerability, how vulnerability distributed in the system, and what are the causes of this distribution. Therefore, on the basis of the study area, it is confirmed by various authors that poverty, weak agrosystems, weak technology, inadequate knowledge, and over dependency on rain-fed agriculture are among the key reasons for increased vulnerability [5, 12, 25, 29, 30]. Rain-fed agriculture that takes over 70% of the population has become a risk business in the area [3].

The situation has destroyed the production systems; thus, even the drought-resistant crops cannot further withstand [31]. Thus far, food shortage and poverty have increased in the region [24, 32–34]. This has mainly happened due to crop failure. In addition, thousands of animals have died due to drought. This has been more pronounced in Tanzania, Kenya, and Ethiopia which experience semiarid climates [35]. The vulnerability has been more pronounced during critical areas due to limited livelihood options [36]. For example, the government can have capacity to help a bearable number of people, however, if the number of people elapses beyond its controlling capacity, it is obviously that the excess number of people will be helpless and thus, conceding intensive impacts.

7. Impacts of climate change

Despite the arguments of some pessimists and lobbyists on climate change, it is obvious that climate impacts and vulnerability have ruined the livelihoods of many people [6]. This has been more pronounced to the already affected agro-ecosystems [4, 24, 31, 34, 37]. In most cases, climate impacts have been measured in different contexts. Farmers do so during crop growing season, while livestock keepers measure it during drought season when pasture and water for their herds are inadequate. Various studies have modeled climate trends to depict the general alterations [5, 12, 22, 32].

Further, Mkonda [38] analyzed the temporal wet spells in Tanzania during the growing season and found that in most years there were seven to nine wet spells in January and March, while below seven wet spells were recorded during February. Recently, the study by Ghebrezgabher et al. [8] had almost similar variation of wet spells when it analyzed the dry and wet condition in the Horn of Africa (Eastern Africa). Similar trend was observed by Stern et al. [39] some decades back. These wet spells are always high in the equatorial climatic region. However, despite the ubiquitous of wet spells in these areas, there have been insignificant impacts to crop production if they are highly unevenly distributed in a particular month because prolonged dry spells may take over to affect crop production. Therefore, this scenario can be curbed by improving the forecasting methods to determine the real climatic situation.

The study by Kahsay and Hansen [9] observed that temperature and rainfall have been varying over space and time in the region (Eastern Africa). It revealed that during spring, the mean rainfall has been kept at 266 mm, while during fall and summer, it has been kept at 201 and 133 mm, respectively. On the other hand, the mean temperature for spring, fall, and summer has been kept at 24.58, 24.02, and 24.28°C, respectively. Our review implies that there has been a modest association between the variation of rainfall and temperature. Therefore, the alteration of these climate variables during the growing seasons has been bringing significant impacts to agricultural output while being largely irrelevant outside the growing season.

Although the impact of climate change can affect both the managed and unmanaged ecosystems and the livelihood of the majority, it rather hits most the vulnerable people as soundly stipulated in Section 4 of this paper. Livestock keepers lack optional livelihoods when thousands of their cattle die due to extreme drought [16, 18]. In due course, the climate affects livestock in a number of ways; an increase in heat affects the health of animals and reduces their food intake. The situation also affects the dairy and meat and production as a result from impacts on grass and rangeland [40]. Apart from that, the situation deprives animals from access to plenty of water due to drought [40, 41]. However, pastoralist with some financial muscles shift from one place to another using transport systems while those with weak economy were severely affected with their herds [14].

For instance, since 2000, Tanzania has experienced massive movements of pastoral communities (especially the Maasai) with their herds in thousands. In most cases the movement has been from areas with low potentials, e.g., semiarid, to high-potential zones, e.g., in floodplains [31], in search for pastures and water. This has been an adaptation strategy of these societies to get rid from extreme droughts. However, this has not been a sustainable option since they have further been degrading the area of destination and, thus, limiting more livelihood options and increasing poverty [28, 41]. IPCC reports have already specified that poor people with less socioeconomic instruments have been the most victims of climate impacts [1, 2]. The study by Agrawala et al. [29] that was conducted in Tanzania had similar observations. This, therefore, shows the degree of magnitudes of climate impacts to vulnerable livelihoods. Despite the direct impacts of climate to various livelihoods and ecosystems, there is a wide range of indirect impacts that also hits the same ecosystems and living organisms. And this poses accumulative impacts to the same.

8. Existing adaptation strategies

According to UN Framework Convention on Climate Change, adaptation and mitigation are very important practices for the sustainability of the planet. Despite the fact that these two practices are credible in both developed and developing countries, IPCC reports [1–3] have affirmed that developing countries, especially the sub-Saharan Africa, need to devote more time and resources to significantly embody these practices into all sectors that determine the peoples' livelihoods. Eastern Africa is among the regions with rapid population increase (growth rate 3%); thus, the demands of food and other environmental resources are increasing rapidly [42]. So far, if the increasing food demand due to human population growth is not well curbed, more degradation or the misuse of environmental assets and natural resources can be enormous. This situation has progressively compelled the region to adapt and cope with the changing climate to meet the necessary environmental resources even though this has been done with little success.

Various studies such as Alqudah et al. [43], Burney and Naylor [44], Cole [45], Eriksen et al. [46], Kilembe et al. [47], Lobell and Gourdji [48], and Rickards and Howden [49] have observed that the existing adaptation practices in Eastern Africa have widely been incremental, or transformational and have been basing on the severity of the climate impacts. Incremental adaptation refers to the fine-tuning of the existing system to minimize impacts, which includes changing planting dates, crop varieties, plant density, and nutrient and water management practices, while transformational involves the opting of alternative income generation methods [49, 50].

The increment adaptation has been the most important aspect as many people remain in agricultural production as their major livelihoods. These adaptations have significantly involved irrigation (i.e., groundwater or from rainwater harvest), fertilization, and adoption of drought and tolerant seeds. This is dominant in the countries like Tanzania, Uganda, Rwanda, and Burundi where agriculture contributes over 50% of the GDP [10]. Even in countries like Kenya and Ethiopia where agriculture contributes to less than 30% of the GDP, adoption of suitable adaptation measures has been unavoidable.

9. Potential trajectories to improve adaptation and mitigation strategies

The current situation shows that despite having numerous strategies, practices, and programs to address climate change impacts, the level of vulnerability is still high in most developing countries [51]. This is because there are other factors that affect negatively the efforts of increasing resilience to climate impacts. Therefore, to improve adaptation to climate impacts, we need to have a clear exploration of what we have been doing for couple of time. In most Eastern African countries, the National Adaptation Plan of Action (NAPA) was adopted in the 2000s [31]. NAPA reviewed the level of vulnerability of economic, social, and ecological factors, among others, to climate change. Thus, it is better to review the NAPA if it still captures the most current challenges. If not, we need to revise it and mainstream the new important aspects in plans, programs, and policies. This will serve the widest audience of victims in the region, more particularly the small-holder farmers.

Again, the study on local condition should be done to explore the actual climate impacts that emanate in the locality. It was found that there have been intra-variations of climate variables within the local condition [12]. This now necessitates the follow-up from that level and progresses onward. However, this should be accompanied by authentic meteorological stations installed in villages. This will be the source of climate data and information.

Besides, various international and local reports have realized that since rain-fed agriculture is not reliable, there is a need to develop irrigation systems from groundwater and rainfall harvest sources [1]. In semiarid zones such as Central Tanzania and Northeastern Kenya and Uganda, intensive transformation to serve the livelihoods in these areas is needed. It will improve the agricultural systems for food crops and animal husbandry. This is because, currently, there has been a massive crop failure and death of a huge number of animals.

The improvement of agricultural systems needs to be given high priority in the region. This is because most poor agricultural systems have significant contribution to the emission of greenhouse gases. This is influenced by limited knowledge and technology to overcome the environmental changes that have been progressing over time. The implementation of various plans, projects, and programs needs to have effectiveness to the majority and not business as usual. This involves the best addressing of the climate challenge to the targeted group. Despite the increasing number of needy people, however, there is still a possibility to advance the methods of solving the challenges as related to the magnitude of the problem. This will bring more positive results to the social, economic, and ecological development to the majority.

10. Conclusions and potential policy implications

In this review, we assess and estimate the causes, vulnerability, impact, and adaptation strategies of climate change on the livelihoods in Eastern Africa. This was to respond the call for the Fifth Assessment Report of the IPCC regarding climate change in developing countries to curb the dominant greenhouse gases for the sustainability of the planet. Our main contribution is to indicate the level of climate impacts by giving evidences from various robust and scientific researches that have been done in favor of the scope of the present study.

Here we find that a wide range of anthropogenic activities in most developing countries involve deforestation, degradation, and pollution of the environment and, thus, emit tons of greenhouse gases (i.e., CO_2, CH_4, N_2O, and CFCs) into the atmosphere. This is the claim of various climate stakeholders especially in international conferences (i.e., COP). Despite being less polluters than their counterpart, i.e., developed countries, the developing countries need to improve and stabilize their adaptation and mitigation measures because they suffer the most and are least equipped to cope. The variation of mean season temperature and precipitation within growing season has had a significant impact to agricultural production in the region. This in turn poses more vulnerability to farmers especially the poor, thus depriving the tool to either heal or cope with the dreadful condition.

Our estimation results appear to be economically viable, environmentally friendly, and communality acceptable as they consider the actual situation of the majority farmers and their socioeconomic dynamics. In addition, they can help to plan, prepare, and implement sound climate policies in the regions and/or international level. This will help in attributing farmer responses to climate variability with respect to socioeconomic and ecological circumstances.

The study has also viewed that there is substantial potential adaptation and mitigating measures of climate change which possibly can even increase agricultural output through conventional technologies such as flexible planting and rainwater harvesting, conservational agriculture (i.e., agroforestry), afforestation, and sustainable utilization of the Earth's resources. It also realized the need to improve forecasting methods at relevant scales for understanding ecosystem response and translating forecasts into useful decision support for natural resource managers and farmers. Therefore, there is a need to document all sound adaptation and mitigation measures that have proved to be helpful in the region. This should go with good governance in the region as some countries have political stiffness due to allegation of being dictatorial regimes.

Practically, adaptation strategies should be significantly mainstreamed into the country's planning frameworks, but how? By conducting vulnerability assessments for critical sectors to enhance understanding of the potential impacts of climate change. Developing a national climate change strategy that clearly lays out priority sector and ecosystem vulnerabilities and means for addressing them. Overall, this will improve the resilience of the people in developing countries and make the planet free from excessive concentration of greenhouse gases that are increasingly disturbing the destiny of the planet.

Conflict of interest

The author declares no conflict of interest.

Author details

Msafiri Yusuph Mkonda
Department of Geography and Environmental Studies, Solomon Mahlangu College of Science and Education, Sokoine University of Agriculture, Morogoro, Tanzania

*Address all correspondence to: msamkonda81@yahoo.co.uk

References

[1] IPCC. Climate Change 2007: Impacts, Adaptation and Vulnerability—Contribution of Working Group II to the Third Assessment Report of the Intergovernmental Panel on Climate Change. Cambridge, United Kingdom/New York, NY, USA: Cambridge University Press; 2007

[2] IPCC. Climate Change 2007: The Physical Science Basis Contribution of Working Group I to the Fourth Assessment Report of the Intergovernmental Panel on Climate Change. Cambridge, United Kingdom/New York, NY, USA: Cambridge University Press; 2007

[3] IPCC. Summary for policymakers. In: Field CB, Barros VR, et al., editors. Climate Change 2014: Impacts, Adaptation, and Vulnerability. Part A: Global and Sectoral Aspects, Contribution of Working Group II to the Fifth Assessment Report of the Intergovernmental Panel on Climate Change. Cambridge, UK: Cambridge University Press; 2014. pp. 1-32

[4] Hens L. An evidence based data set on climate changes for developing countries. Environment, Development and Sustainability. 2014;**16**(2):255-256. DOI: 10.1007/s10668-013-9504-7

[5] Ahmed S, Deffenbaugh N, Hertel T, Lobell D, Ramankutty N, Rios A, et al. Climate volatility and poverty vulnerability in Tanzania. Global Environmental Change. 2011;**21**(2011):46-55

[6] Blanc E. The impact of climate change on crop yields in sub-Saharan Africa. American Journal of Climate Change. 2012;**1**:1-13

[7] Hertel T, Burke M, Lobell D. The poverty implications of climate-induced crop yield changes by 2030. Global Environmental Change. 2010;**20**:577-585

[8] Ghebrezgabher M, Yang T, Yang X. Long-term trend of climate change and drought assessment in the horn of Africa. Advances in Meteorology. 2016;**2016**:12. Article ID: 8057641. DOI: 10.1155/2016/8057641

[9] Kahsay G, Hansen L. The effect of climate change and adaptation policy on agricultural production in eastern Africa. Ecological Economics. 2016;**121**:54-64

[10] FAOSTAT. FAO Statistical Database. 2014. Available from: http://faostat3.fao.org/home/E [Accessed: March 2017]

[11] Eriksen S, Noess L. Pro-Poor Climate Adaptation: Norwegian Development Cooperation and Climate Change Adaptation. Oslo: Norwegian Agency for Development Cooperation; 2003

[12] Rowhani P, Lobell DB, Linderman M, Ramankutty N. Climate variability and crop production in Tanzania. Agricultural and Forest Meteorology. 2011;**151**(2011):449-460

[13] Kleinman P, Bryant R, Pimentel D. Assessing ecological sustainability of slash-and-burn agriculture through soil fertility indicators. Agronomy Journal. 1996;**88**(2):122-127. DOI: 10.2134/agronj1996.00021962008800020002x

[14] Sen A. Food, Economics and Entitlements. WIDER Working Papers (1986-2000) 1986/001. Helsinki: UNU-WIDER; 1986

[15] Maslin M, Brierley C, Milner A, Shultz S, Trauth M, Wilson K. East African climate pulses and early human evolution. Quaternary Science Reviews. 2014;**101**:1-17

[16] Nyssen J, Poesen J, Moeyersons J, Deckers J, Haile M, Lang A. Human impact on the environment in the

Ethiopian and Eritrean highlands—A state of the art. Earth-Science Reviews. 2004;**64**(3-4):273-320

[17] Kiage L, Liu K. Palynological evidence of climate change and land degradation in the Lake Baringo area, Kenya, East Africa, since AD 1650. Palaeogeography, Palaeoclimatology, Palaeoecology. 2009;**279**(1-2):60-72

[18] Pricope N, Husak G, Lopez-Carr D, Funk C, Michaelsen J. The climate-population nexus in the east African horn: Emerging degradation trends in rangeland and pastoral livelihood zones. Global Environmental Change. 2013;**23**(6):1525-1541

[19] Romahn S, Mackensen A, Kuhlmann H, P¨atzold J. Benthic foraminiferal response to late glacial and Holocene Sea level rise and rainfall variability off East Africa. Marine Micropaleontology. 2015;**119**:34-48

[20] Frankl A, Poesen J, Haile M, Deckers J, Nyssen J. Quantifying long-term changes in gully networks and volumes in dryland environments: The case of northern Ethiopia. Geomorphology. 2013;**201**:254-263

[21] FAO. FAO-Geonetwork. Rome, Italy: Food and Agriculture Organization of the United Nations; 2002

[22] Lobell DB, Burke MB. Why are agricultural impacts of climate change so uncertain? The importance of temperature relative to precipitation. Environmental Research Letters. 2008;**3**:034007

[23] Eriksen S, O'Brien K, Rosentrater L. Climate change in eastern and southern Africa: Impacts, vulnerability and adaptation. In: Global Environmental Change and Human Security, Report. Vol. 2008. 2008. p. 2

[24] URT. Review of food and agricultural policies in the United Republic of Tanzania. In: MAFAP Country Report Series. Rome, Italy: FAO; 2014

[25] Adger N. Vulnerability, global environmental change. Journal of Environment. 2006;**16**:268-281

[26] Mahoo HF, Young MDB, Mzirai OB. Rainfall variability and its implications for the transferability of experimental results in the semi-arid areas of Tanzania. Special issue: Rain water harvesting for crop production in semi-arid Tanzania. Tanzania Journal of Agricultural Sciences. 1999;**2**(2):127-140

[27] Baumann F. Improving Access to Natural Resources for the Rural Poor: A Critical Analysis of Central Concepts and Emerging Trends from a Sustainable Livelihoods Perspectives. Working Paper I. Rome: Livelihood Support Programme; 2002

[28] Crane T, Delaney A, Tamás P, Chesterman S, Ericksen P. A Systematic Review of Local Vulnerability to Climate Change in Developing Country Agriculture; 2007. DOI: 10.1002/wcc.464

[29] Agrawala SA, Moehner A, Hemp M, Van Aalst S, Hitz J, Smith H, et al. Development and Climate Change in Tanzania: Focus on Kilimanjaro. Paris: Organisation for Economic Co-operation and Development; 2003

[30] Burke M, Lobell D. Food security and adaptation to climate change: What do we know? Climate Change and Food Security. 2009;**37**:133-153

[31] URT. United Republic of Tanzania, National Adaptation Programme of Action (NAPA). Dar es Salaam: Division of Environment, Vice President's Office; 2007

[32] Lobell DB, Burke MB. On the use of statistical models to predict crop

yield responses to climate change. Agricultural and Forest Meteorology. 2010;**150**:1443-1452

[33] Mbilinyi BP, Tumbo SD, Mahoo HF, Senkondo EM, Hatibu N. Indigenous knowledge as decision support tool in rainwater harvesting. Physics and Chemistry of the Earth, Parts A/B/C. 2005;**30**:792-798

[34] URT. Poverty and Human Development Report Research and Analysis Working Group; 2009

[35] Mkonda MY, He XH. Are rainfall and temperature really changing? Farmer's perceptions, meteorological data, and policy implications in the Tanzanian semi-arid zone. Sustainability. 2017;**9**(8):1412. DOI: 10.3390/su9081412

[36] Mkonda MY, He XH, Festin ES. Comparing smallholder farmers' perception of climate change with meteorological data: Experiences from seven agro-ecological zones of Tanzania. Weather, Climate, and Society. 2018;**10**(3):435-452. DOI: 10.1175/WCAS-D-17-0036.1

[37] Mkonda MY, He XH. Yields of the major food crops: Implications to food security and policy in Tanzania's semi-arid agro-ecological zone. Sustainability. 2017;**9**(8):1490. DOI: 10.3390/su9081490

[38] Mkonda MY. Temporal rainfall and temperature trends, impacts in agriculture and adaptations that respond to local conditions. A case of Mvomero District, Tanzania. Research on Humanities and Social Sciences. 2014;**4**(12):36-49

[39] Stern RD, Dennett MD, Dale IC. Analyzing rainfall measurements to give agronomically useful results. II. Modelling approach. Experimental Agriculture. 1982;**18**:237-253

[40] Tietjen B, Jeltsch F. Semi-arid grazing systems and climate change: A survey of present modelling potential and future needs. Journal of Applied Ecology. 2007;**44**:425-434

[41] Thornton P, Herrero M, Freeman A, Mwai O, Rege E, Jones P, et al. Vulnerability, Climate Change and Livestock—Research Opportunities and Challenges for Poverty Alleviation. Kenya: ILRI; 2008

[42] FAO. Adapting to Climate Change through Land and Water Management in Eastern Africa: Results of Pilot Projects in Ethiopia, Kenya and Tanzania. Rome: Food and Agricultural Organization of the United Nations; 2014

[43] Alqudah AM, Samarah NH, Mullen RE. Drought stress effect on crop pollination, seed set, yield and quality. In: Lichtfouse E, editor. Alternative Farming Systems, Biotechnology, Drought Stress and Ecological Fertilization. Dordrecht, The Netherlands: Springer; 2011. pp. 193-213

[44] Burney JA, Naylor RL. Smallholder irrigation as a poverty alleviation tool in sub-Saharan Africa. World Development. 2012;**40**:110-123

[45] Cole M. Rwanda's Climate: Observations and Projections. Oxford, United Kingdom: Smith School of Enterprise and the Environment, University of Oxford; 2011

[46] Eriksen S, Klein RJT, Ulsrud K, Næss LO, Brien KO. Climate Change Adaptation and Poverty Reduction: Key Interactions and Critical Measures. GECHS Report 2007:1. Norway: Global Environmental Change and Human Security, University of Oslo; 2007

[47] Kilembe C, Thomas TS, Waithaka M, Kyotalimye M, Tumbo S. East African Agriculture and

Climate Change: A Comprehensive Analysis-Tanzania. Washington, DC: International Food Policy Research Institute; 2012

[48] Lobell DB, Gourdji SM. The influence of climate change on global crop productivity. Plant Physiology. 2012;**160**:1686-1697

[49] Rickards L, Howden SM. Transformational adaptation: Agriculture and climate change. Crop & Pasture Science. 2012;**63**:240-250

[50] Paavola J. Livelihoods, vulnerability and adaptation to climate change in Morogoro, Tanzania. Environmental Science & Policy. 2008;**11**:642-654

[51] Mkonda MY, He XH. Climate variability, crop yields and ecosystems synergies in Tanzania's semi-arid agro-ecological zone. Ecosystem Health and Sustainability. 2018;**4**(3):1-14. DOI: 10.1080/20964129.2018.1459868

Chapter 4

WMO Space-Based Weather and Climate Extremes Monitoring Demonstration Project (SEMDP): First Outcomes of Regional Cooperation on Drought and Heavy Precipitation Monitoring for Australia and Southeast Asia

Yuriy Kuleshov, Toshiyuki Kurino, Takuji Kubota, Tomoko Tashima and Pingping Xie

Abstract

To improve monitoring of extreme weather and climate events from space, the World Meteorological Organization (WMO) initiated the space-based weather and climate extremes monitoring demonstration project (SEMDP). Presently, SEMDP is focused on drought and heavy precipitation monitoring over Southeast Asia and the Pacific. Space-based data and derived products form critical part of meteorological services' operations for weather monitoring; however, satellite products are still not fully utilized for climate applications. Using SEMDP satellite-derived precipitation products, it would be possible to monitor extreme precipitation events with uniform spatial coverage and over various time periods – pentad, weekly, 10 days, monthly and longer time-scales. In this chapter, SEMDP satellite-derived precipitation products over the Asia-Pacific region produced by the Earth Observation Research Center/Japan Aerospace Exploration Agency (EORC/JAXA) and the Climate Prediction Center/National Oceanic and Atmospheric Administration (CPC/NOAA) are introduced. Case studies for monitoring (i) drought in Australia in July-October 2007 and September 2018 and (ii) heavy precipitation over Australia in December 2010 and Thailand and the Peninsular Malaysia in November-December 2014 which caused widespread flooding are also presented. Satellite observations are compared with in situ data to demonstrate value of satellite-derived estimates of precipitation for drought and heavy rainfall monitoring.

Keywords: weather and climate extremes, drought, heavy rainfall, space-based observations, Asia-Pacific

1. Introduction to SEMDP

Meteorological observations clearly demonstrate that the global climate change occurs since the beginning of the industrial revolution, with particular rapid change since about 1950, including changes in weather and climate extreme events [1]. This increase in weather and climate extremes leads to significant increase in impact of natural disasters on society worldwide. One of the world's most disaster-prone regions is Asia-Pacific. In this region, almost 2 million people were killed in disasters between 1970 and 2011, representing 75% of all disaster fatalities globally; the most frequent hazards in the region are hydrometeorological [2]. The increase in frequency and severity of weather and climate extreme events and their impact on society requires the development and implementation of new tools for monitoring these hazardous phenomena globally using modern satellite remote sensing techniques.

Recognizing the importance of this issue, in February 2017 the World Meteorological Organization organized a workshop on operational space-based weather and climate extremes monitoring demonstration project which was attended by representatives of satellite operators, research and development space agencies, Regional Climate Centres (RCCs), and National Meteorological and Hydrological Services (NMHSs) to stimulate a dialog about enhancing utilization of space-based observation data and products for monitoring weather and climate extremes.

The workshop recognized that significant progress has been made in recent years in developing space-based observations in most geophysical fields and that several high-resolution satellite products were available on a quasi-real-time basis, enabling enhanced utilization for monitoring weather and climate extremes from space. It was also recognized that for many developing and least developed countries, strengthening human and technological capacity is required to provide an adequate level of services. As such, transfer of knowledge from countries with greater technological developments is essential, in order to fully utilize advantages of modern space-based data and derived products in developing countries.

Following the workshop's recommendations, WMO initiated SEMDP – the space-based weather and climate extremes monitoring demonstration project. SEMDP is established to run initially for 2 years (2018–2019) and be focused on weather and climate extremes such as drought and heavy precipitation over the Southeast Asia region and the Pacific Ocean. Space-based data and derived products form critical part of operations at NMHSs and RCCs for weather monitoring; however, satellite products are not fully utilized yet for climate applications.

Most NMHSs in countries of Southeast Asia and the Pacific use conventional surface-based rain gauge observations for extreme precipitation monitoring. Rain gauge observations provide accurate measurements of precipitation; however, data are restricted to locations of meteorological observation stations. For example, spatial distribution of rain gauges over Australia is not uniform: while eastern and southern parts of the country, southwest of Western Australia and northern and eastern parts of Tasmania, are densely covered by observation stations, spatial coverage of interior parts of Australia is poor. This issue of nonuniform spatial coverage is typical for countries in the Asia-Pacific region, and the density of rain gauges in many areas is considered as inadequate by users. In contrast with conventional surface-based observations, rainfall estimates derived from global space-based observations better address users' needs for precipitation information providing uniform spatial coverage.

The satellite-based rainfall estimates are based on retrieval algorithms of passive instrumental measurements (radiometry) relating radio signals recorded in infrared and microwave bands of electromagnetic spectrum to the occurrence and intensity of precipitation. Infrared instruments record signals around 11 μm wavelength providing information about cloud top temperature and then applying mathematical retrieval algorithms converting it to estimates of precipitation. Microwave instruments utilize a broad range of electromagnetic spectrum from 10 to 100 GHz. Channels up to 37 GHz primarily provide information about liquid precipitation in the lower parts of clouds; retrieval algorithms are based on assumption that larger amounts of liquid emit higher amount of microwave radiation. Radio signals received through channels above 37 GHz are primarily used for precipitation estimates in the upper parts of clouds due to scattering microwave radiation by solid precipitation. Microwave satellite-borne instruments are employed on the Tropical Rainfall Measuring Mission (TRMM) and the Global Precipitation Measurement (GPM) mission.

Space-based observations can also address users' needs for information about precipitation extremes on short time scales. Current operational climate products for drought monitoring derived from surface-based observations are typically focused on identifying rainfall deficits over extended periods (months to years) using percentile and/or decile analysis. As for heavy precipitation, they are typically diagnosed on a monthly time scale. Using space-based observations, it would be possible to monitor extreme precipitation events over shorter time periods—pentad (5 days), week (7 days), and longer periods of up to a month—in order to respond to current and future users' requirements. Monitoring weather and climate extremes on shorter time scales is considered by RCCs and NMHSs as a valuable extension of their operational products to enhance climate services for users in Asia-Pacific.

In this chapter, WMO SEMDP and its implementation strategy are described, and first outcomes of Asia-Pacific regional cooperation on drought and heavy precipitation monitoring from space are presented.

2. SEMDP precipitation products

SEMDP is designed as a demonstration project to bring benefits of utilizing space-based observations of extreme precipitation to operational services of RCCs and NMHSs. During the project's first implementation stage, SEMDP's geographical domain covers the Southeast Asia region and the Pacific Ocean—area from 40°N to 45°S and 50°E to 160°W. Two agencies—the Earth Observation Research Center/Japan Aerospace Exploration Agency and the Climate Prediction Center/National Oceanic and Atmospheric Administration—provide satellite data and products for the SEMDP region. It is planned to (i) gradually expand SEMDP's geographical domain during subsequent stages of the project's implementation to accomplish the global coverage for SEMDP products and (ii) involve more space and meteorological agencies from around the world to contribute to providing RCCs and NMHSs with a range of SEMDP products.

SEMDP is focused on monitoring extreme events. "Extreme weather event" and "extreme climate event" according to the IPCC AR5 WG I report are defined as follows. "An extreme weather event is an event that is rare at a particular place and time of year. Definitions of rare vary, but an extreme weather event would normally be as rare as or rarer than the 10th or 90th percentile of a probability density function estimated from observations. By definition, the characteristics of what is called extreme weather may vary from place to place in an absolute sense. When a pattern of extreme weather persists for some time, such as a season, it may be classed as an extreme climate event, especially if it yields an average or total that is itself extreme (e.g., drought or heavy rainfall over a season)." [3]. Thus, SEMDP approach to defining drought and heavy rainfall is based on the above definitions.

Based on the workshop's recommendations and consequent consultations with the satellite data providers (EORC/JAXA and CPC/NOAA) and users (RCCs and NMHSs in Southeast Asia and the Pacific), SEMDP aims to satisfy users' requirements for monitoring precipitation extremes on short time scales, i.e., on pentad (5 days) to weekly up to monthly basis utilizing satellite-based products available on near real-time basis for monitoring "heavy precipitation" and "drought" events on a routine basis ("operationally") for climate analysis and monitoring and for the development of improved climate services. Brief introduction of EORC/JAXA and CPC/NOAA satellite-derived data and products available for RCCs and NMHSs in Asia-Pacific is given below.

SEMDP precipitation products produced by EORC/JAXA are based on the Global Satellite Mapping of Precipitation (GSMaP) [4]. GSMaP products are in high demand—more than 4200 users from 114 countries from around the world are registered for the GSMaP data distribution. For SEMDP users in Asia-Pacific, EORC/JAXA provides mean precipitation estimates derived from GSMaP version 6 for hourly, daily (00–23 UTC), pentad (5 days), weekly (Monday–Sunday), 10-day, and monthly precipitation with spatial resolution of 0.1°lat/lon grid box (an example of monthly precipitation for July is given in **Figure 1**). In addition, statistics for daily, pentad, and weekly extreme precipitation (90th–99th percentiles) and percentage of rainy (≥1 mm/day) days in a month is provided (examples are presented in **Figures 2** and **3**). For drought monitoring, the standardized precipitation index (SPI; 1 month, 2 months, and 3 months) for grid boxes over land with spatial resolution of 0.25°lat/lon grid box is provided.

CPC/NOAA provides SEMDP users with a similar set of products using the Climate Prediction Center morphing technique (CMORPH) satellite precipitation

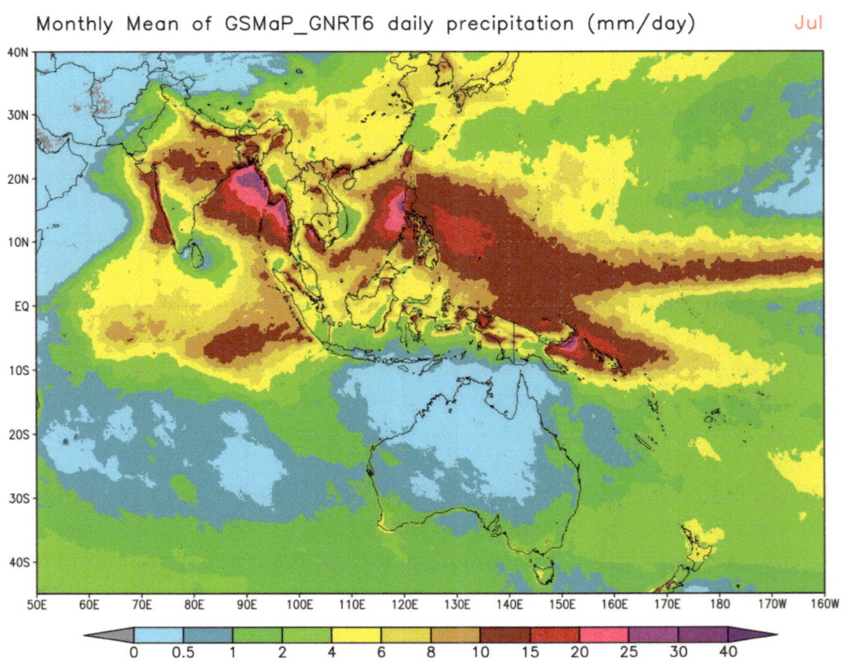

Figure 1.
EORC/JAXA GSMaP monthly mean of daily precipitation for the month of July.

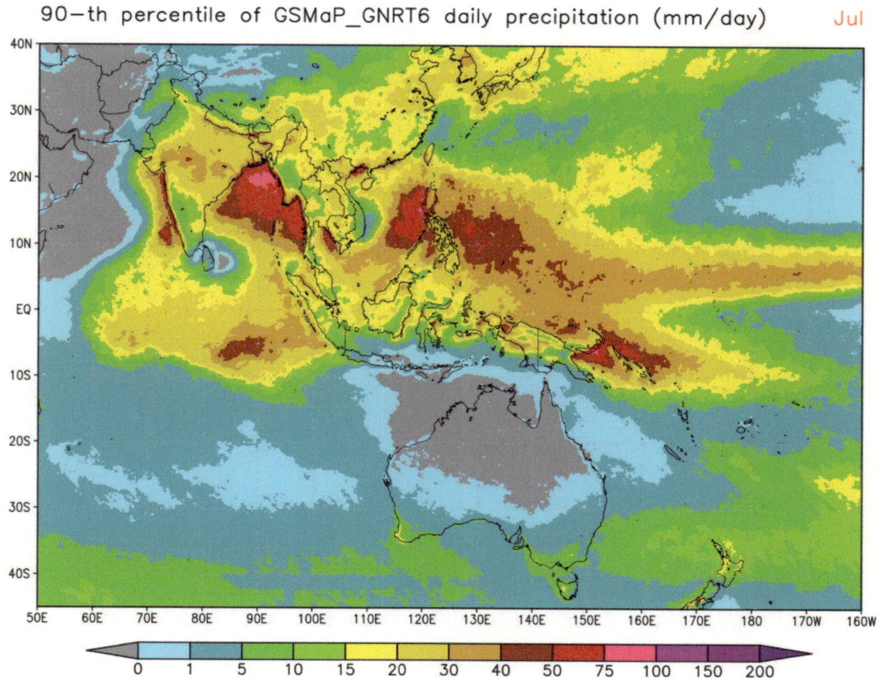

Figure 2.
EORC/JAXA GSMaP 90th percentile of daily precipitation for the month of July.

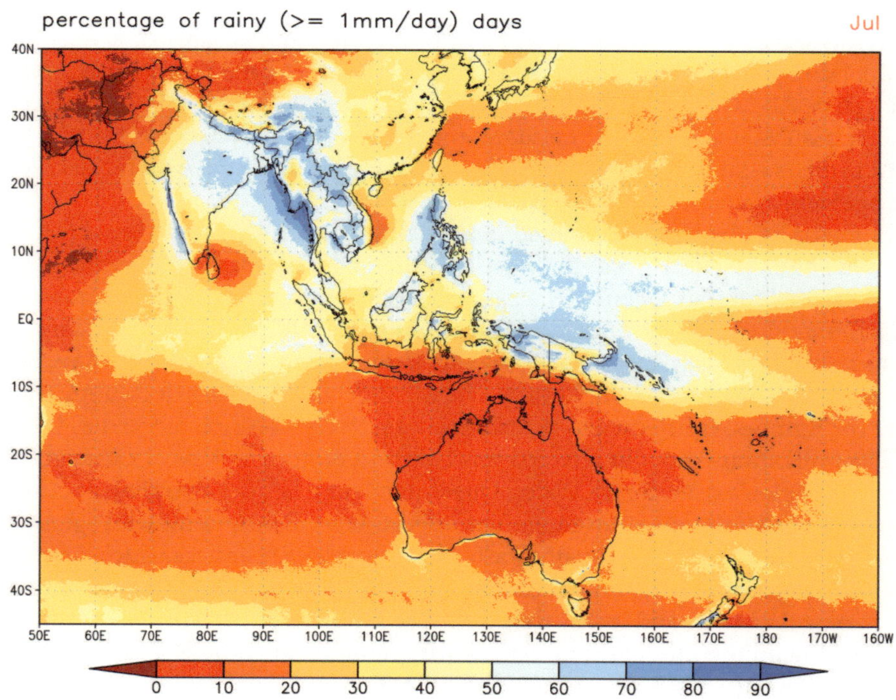

Figure 3.
EORC/JAXA GSMaP percentage of rainy days for the month of July.

estimates (see [5] for detail). In addition to the SPI, weekly normalized differential vegetation index (NDVI; **Figure 4**) and the vegetation health index (VHI) are also available for SEMDP region.

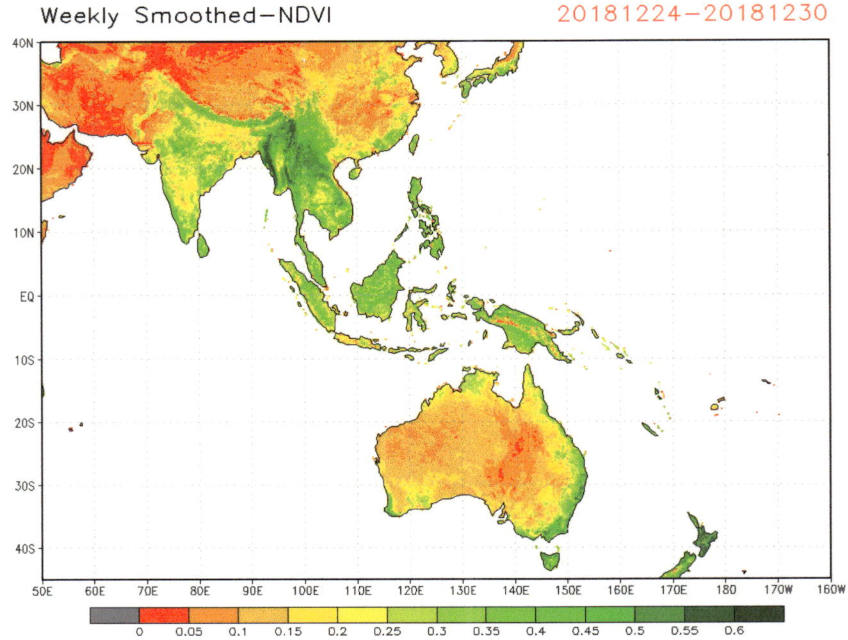

Figure 4.
CPC/NOAA CMORPH weekly NDVI for 24–30 December 2018.

3. Drought monitoring using SEMDP products

In this section, case studies for drought monitoring in Australia using SEMDP products are presented. Australia is the driest continent on the Earth, apart from Antarctica. About 70% of Australia receives less than 500 mm of rain annually, which classifies those parts of the continent as arid or semiarid areas. Drought monitoring is vital for informed decision-making in agriculture, disaster risk management, water management, and other sectors.

In Australia, the Bureau of Meteorology defines drought in the affected region when the rainfall over a 3-month period is being in the lowest decile of what has been recorded for that region in the past [6]. Drought often affects Australia—rainfall observations which the Bureau of Meteorology conducts since the middle of the nineteenth century show that on average drought occurs once every 18 years; severity and duration of drought vary. The worst drought which affected Australia since the European settlement—the Millennium drought—occurred in the 2000s.

The Millennium drought affected southern and eastern regions of the continent (states of Victoria, New South Wales, Queensland, and South Australia), southwest of Western Australia, and Tasmania. The largest Australian agricultural region—the Murray-Darling basin—was severely affected, and water resources which supply cities and towns including capital cities of Melbourne, Sydney, Brisbane, Adelaide, and many other cities and towns were also severely affected.

The Millennium drought commenced with rainfall deficit in 1996–1997 and continued during very dry years in 2001–2002 (**Figure 5**); it was clear that this is the worst drought in Australia on record [7].

During the year 2006 southeastern parts of Australia had the second driest year on record [8]; agricultural region of the Murray-Darling basin was particularly severely affected by drought conditions (**Figure 6**). Drought continued to affect the Murray-Darling basin in 2007; it was already seventh consecutive year of below average rainfall for the basin. Dry and hot conditions continued to affect Australia through to early 2010.

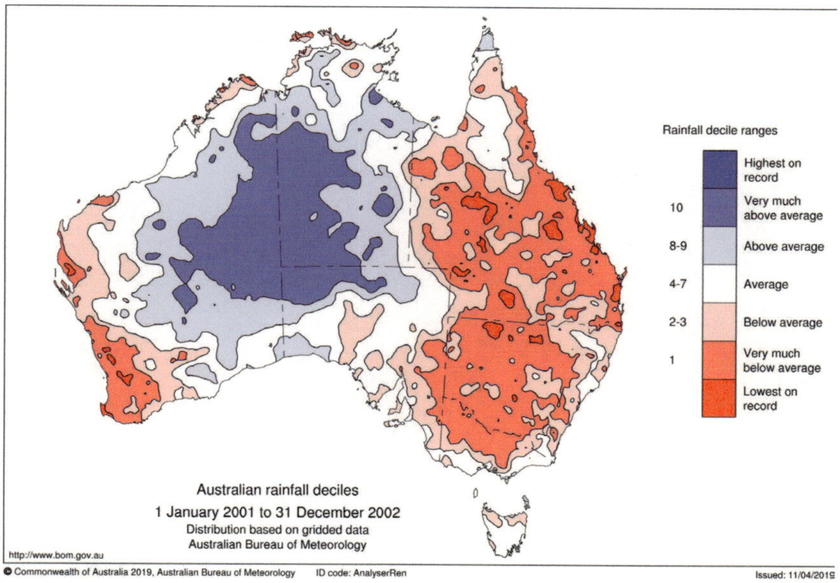

Figure 5.
Rainfall deciles for Australia in January 2001–December 2002 derived from the Australian Bureau of Meteorology rain gauge observations.

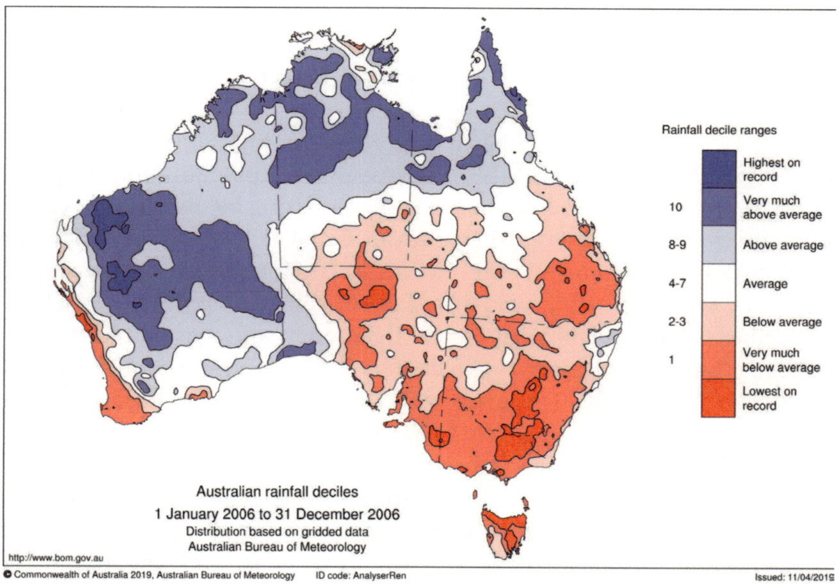

Figure 6.
Rainfall deciles for Australia in January–December 2006 derived from the Australian Bureau of Meteorology rain gauge observations.

The 2010–2011 La Niña event brought the Millennium drought to the end. This La Niña event was one of the strongest on records, and it resulted in record-breaking rainfall in the Murray-Darling basin and above average rainfall over the southeast parts of the country (**Figure 7**). Significant increase in surface water storage and soil moisture due to continuing above average rainfall ended drought conditions in the southeastern parts of Australia [9].

The Millennium drought was the most severe drought which affected Australia over the past few centuries. It is pertinent to examine the usefulness of space-based observations for drought monitoring over Australia; here

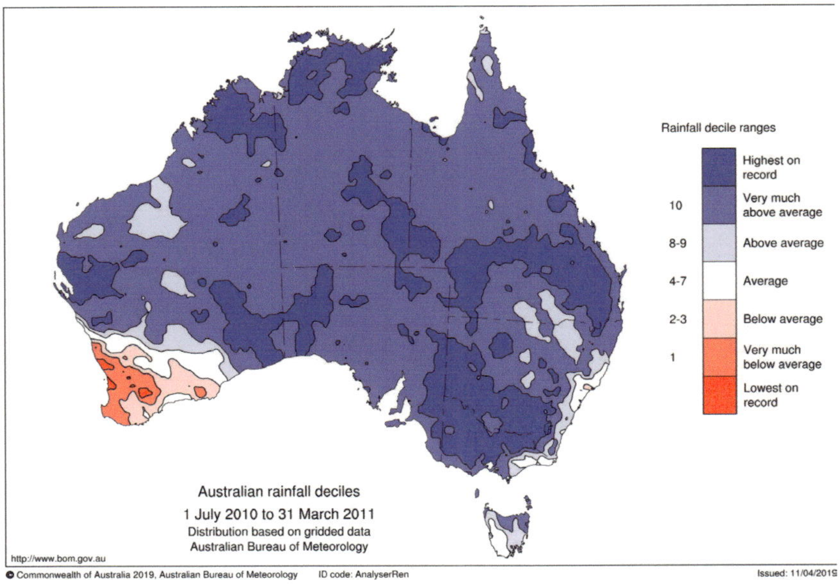

Figure 7.
Rainfall deciles for Australia in July 2010–March 2011 derived from the Australian Bureau of Meteorology rain gauge observations.

we present a case study for the year 2007 of the Millennium drought utilizing rainfall percentile, 1-month and 3-month SPI values derived from the EORC/JAXA GSMaP data.

The SPI is an index which is widely used for meteorological drought detection and monitoring. Positive values of the SPI correspond to precipitation above median, and negative values of the SPI correspond to precipitation below median. Drought conditions are classified when the SPI values are equal to or below −1.0. Specifically, for the SPI values −1.0 and below conditions are classified as "moderately dry," for −1.5 and below as "severely dry," and for −2.0 and below as "extremely dry."

As described above, the main agricultural region in southeastern Australia—the Murry-Darling basin—was severely affected by the Millennium drought. Examining 1-month SPI for August 2007 (**Figure 8**) and rainfall percentile (**Figure 9**) derived from EORC/JAXA GSMaP, one can find that drought-affected areas where the SPI values are less than −1.5 (i.e., "severely dry") correspond well to areas of rainfall below the 10th percentile. The detected by space-based observations drought-affected areas are in good correspondence with areas defined as "very much below average" on rainfall decile map for August 2007 derived from the Australian Bureau of Meteorology rain gauge observations (**Figure 10**). Similarly, areas where values of 3-month SPI for July–September 2007 (**Figure 11**) are below −1.5 are in good correspondence with areas of "very much below average" rainfall on rainfall decile map (**Figure 12**).

It should be noted that space-based and in situ observations are in good agreement over the Murry-Darling basin in southeastern Australia where the density of surface-based observations is high; however, there are noticeable discrepancies between them over the central parts of the continent where the density of surface-based observations is very low. It clearly demonstrates value of space-based rainfall estimates for drought detection and monitoring, especially for regions where rain gauge observations are limited or unavailable.

SEMDP products became available to NMHSs and RCCs in Asia-Pacific on a quasi-operational basis from December 2018, thanks to the dedicated efforts of

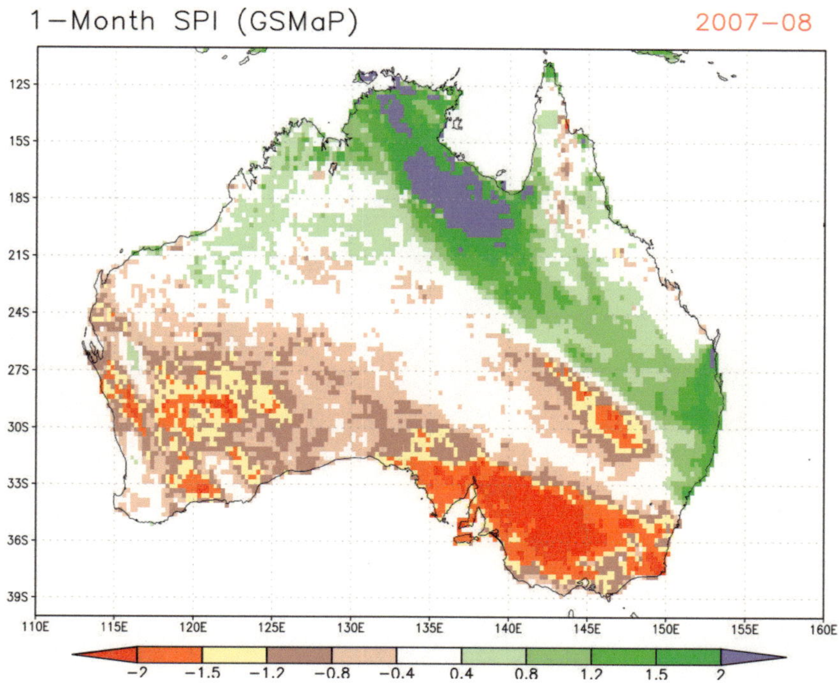

Figure 8.
SPI for Australia in August 2007 derived from the EORC/JAXA GSMaP data.

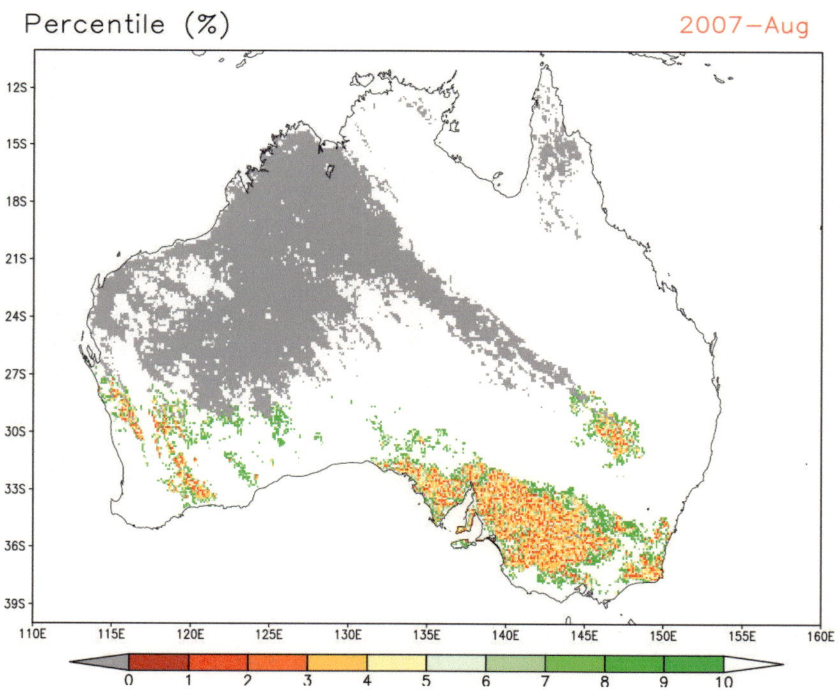

Figure 9.
Rainfall percentiles for Australia in August 2007 derived from the EORC/JAXA GSMaP data.

scientists and IT experts from the EORC/JAXA and the CPC/NOAA. Here we demonstrate usefulness of available SEMDP products for operational drought monitoring in Australia using the VHI. 2018 for Australia was a year of persistent warmth (the third warmest year on record with mean temperature 1.14°C above the

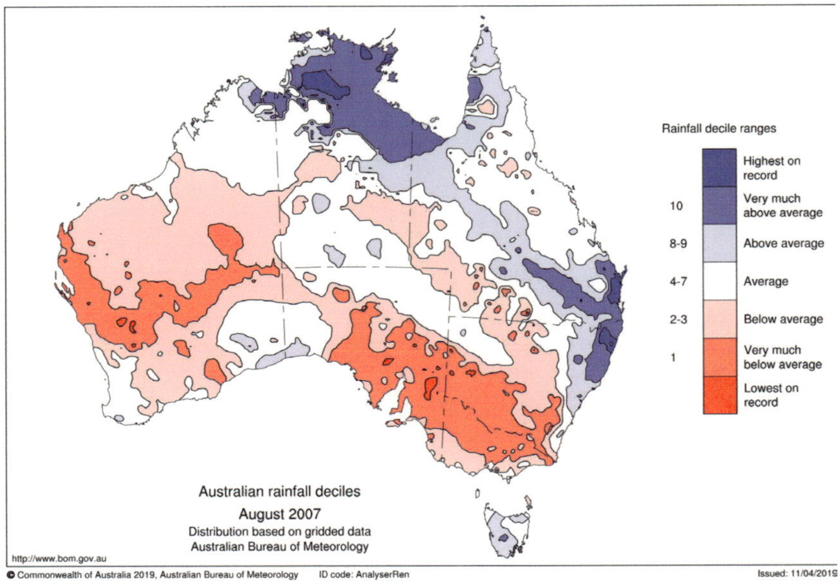

Figure 10.
Rainfall deciles for Australia in August 2007 derived from the Australian Bureau of Meteorology rain gauge observations.

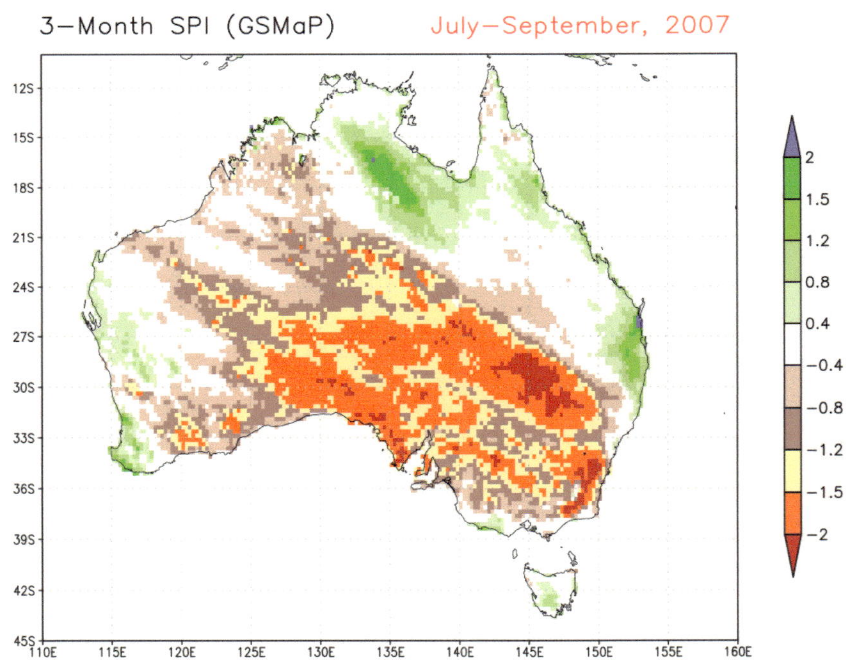

Figure 11.
SPI for Australia for 3 months (July–September 2007) derived from the EORC/JAXA GSMaP data.

1961–1990 average) and protracted drought (average rainfall was 412.8 mm which is 11% below the 1961–1990 average of 465 mm) [10]. Annual rainfall was very low ("very much below average" and "lowest on record") over the southeastern parts of the country and above average in the area between the northwest and southeast of Western Australia (**Figure 13**). Rainfall was particularly low over the southeast from April; September was record-dry. Dry conditions had an impact on vegetation which could be estimated by the vegetation health index.

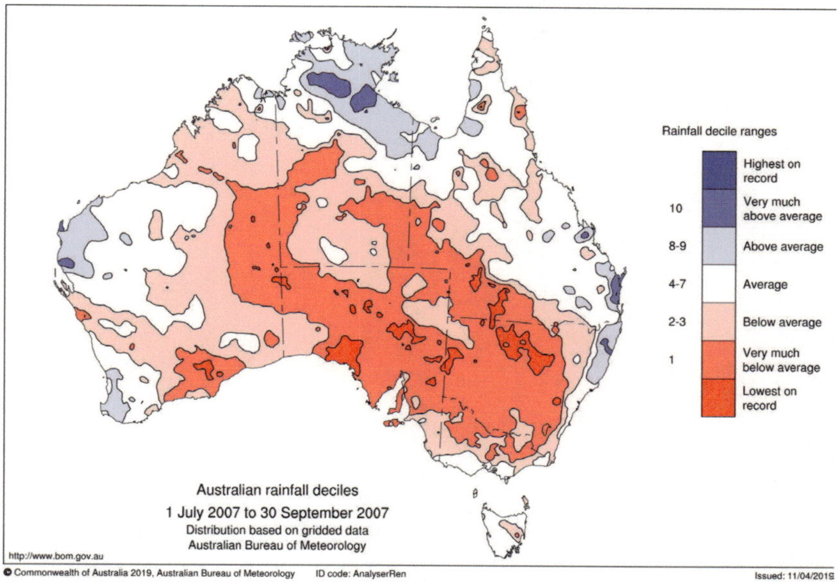

Figure 12.
Rainfall deciles for Australia in July–September 2007 derived from the Australian Bureau of Meteorology rain gauge observations.

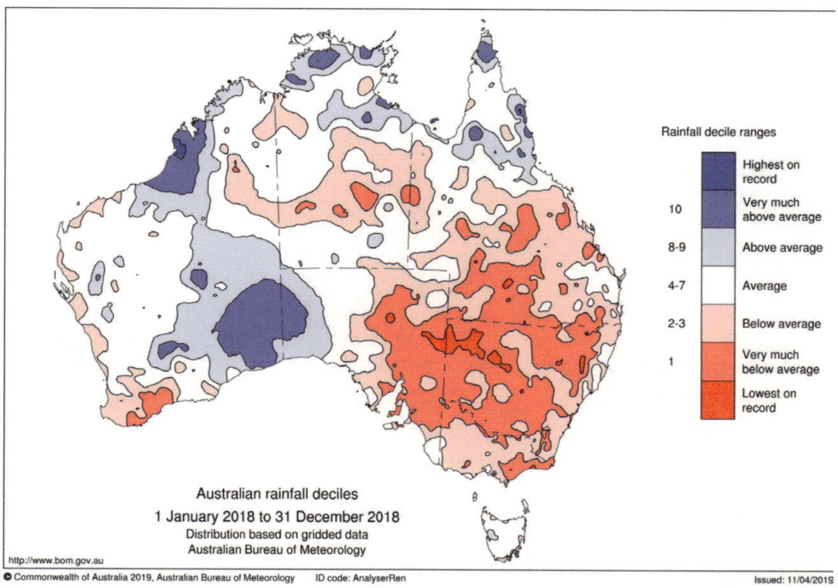

Figure 13.
Rainfall deciles for Australia in January–December 2018 derived from the Australian Bureau of Meteorology rain gauge observations.

The VHI is computed using observations from Advanced Very High Resolution Radiometer (AVHRR) instrument onboard the NOAA polar orbiting satellites in the visible, infrared, and near-infrared bands and used to identify stress on vegetation related to drought [11]. Maps of the VHI for the last week of September 2017 (above average rainfall for Australia was observed in that year) and the last week of September 2018 are presented in **Figure 14** demonstrating difference between relatively healthy vegetation over Australia in September 2017 (**Figure 14a**) and stressed vegetation in September 2018 (**Figure 14b**) due to impact of dry conditions.

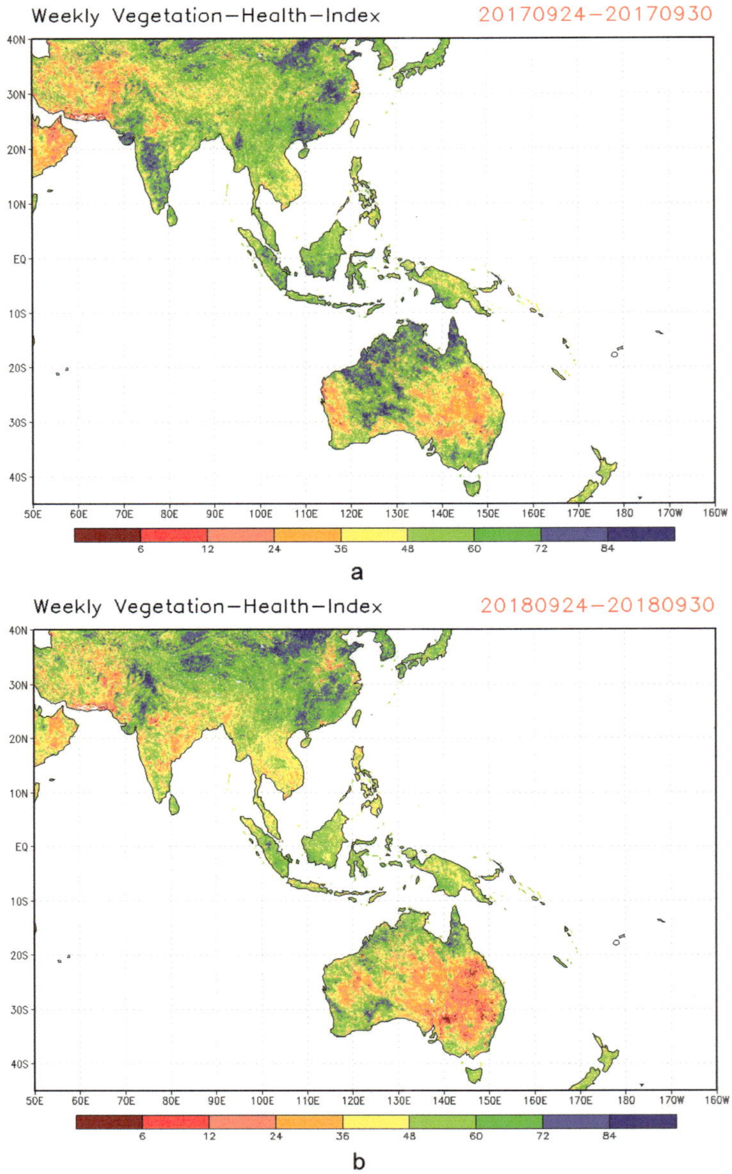

Figure 14.
CPC/NOAA VHI for (a) 24–30 September 2017 and (b) 24–30 September 2018.

4. Heavy precipitation monitoring using SEMDP products

In this section, case studies of heavy precipitation over Australia in December 2010 and Thailand and Peninsular Malaysia in November–December 2014 which caused widespread flooding are presented.

An "extreme rainfall" is defined when a mean rainfall for a specified period is higher than a certain percentile threshold, e.g., 90th–99th percentile (**Figure 15**).

Extreme rainfall associated with La Niña event has been observed over Australia in 2010 and 2011. In 2011, Australia experienced its third wettest year since national rainfall records began in 1900 [12]. Averaged across Australia, both years experienced rainfall well above average—690 mm (225 mm above the long-term average of 465 mm) in 2010 [9] and 699 mm (234 mm above the long-term average of 465 mm) in 2011 [12].

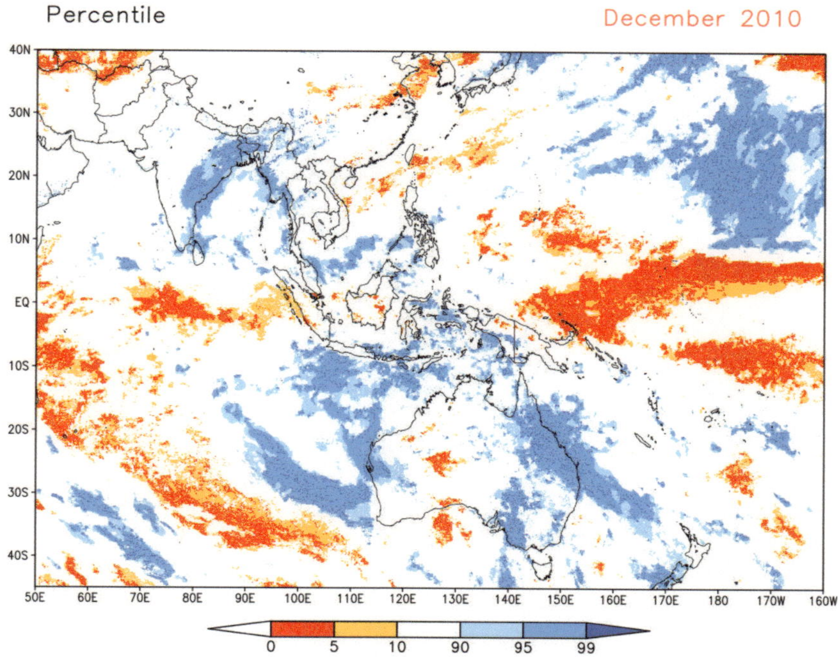

Figure 15.
EORC/JAXA GSMaP rainfall percentile over SEMDP domain for December 2010.

The 2010–2011 La Niña event was one of the strongest on record, comparable in strength with the La Niña events of 1917–1918, 1955–1956, and 1975–1976, and it has significant impact on Australian rainfall. La Niña is typically associated with increased rainfall in northern and eastern Australia. During the 2010–2011 La Niña, most of the mainland Australia experienced significantly higher than average rainfall over the 9 months from July 2010 to March 2011 (**Figure 7**). A number of new Australian rainfall records were set: wettest September, December, and March on record and second wettest October and February. Extreme rainfall associated with La Niña event has been observed over parts of western and eastern Australia in December 2010 (**Figure 16**). The record-breaking rainfall during the 2010–2011 La Niña led to widespread flooding in many regions between September 2010 and March 2011 including southeast Queensland, large areas of northern and western Victoria, New South Wales, northwestern Western Australia, and eastern Tasmania that were subject to significant flooding.

In **Figure 17**, EORC/JAXA GSMaP rainfall percentile over Australia for December 2010 is presented. An area above 95th percentile derived from GSMaP approximately corresponds to an area of rainfall deciles "very much above average" as derived from rain gauge observations by the Australian Bureau of Meteorology (**Figure 16**); it demonstrates that this extreme rainfall event was well detected using GSMaP.

The second case study examines episodes of heavy precipitation over Thailand and Peninsular Malaysia in November–December 2014 [13]. In November 2014, an episode of heavy rainfall and subsequent flooding in the coastal area of northeastern Peninsular Malaysia occurred from 13 to 20 November 2014. In the second half of December 2014, two episodes of heavy precipitation caused widespread flooding in south of Thailand, Kelantan, Terengganu, and Pahang and on the east coast of Peninsular Malaysia.

Accumulated rainfall over Peninsular Malaysia in November and December 2014 derived from GSMaP is presented in **Figure 18a** and **b**, respectively. Time series of daily precipitation for November–December 2014 averaged over land in

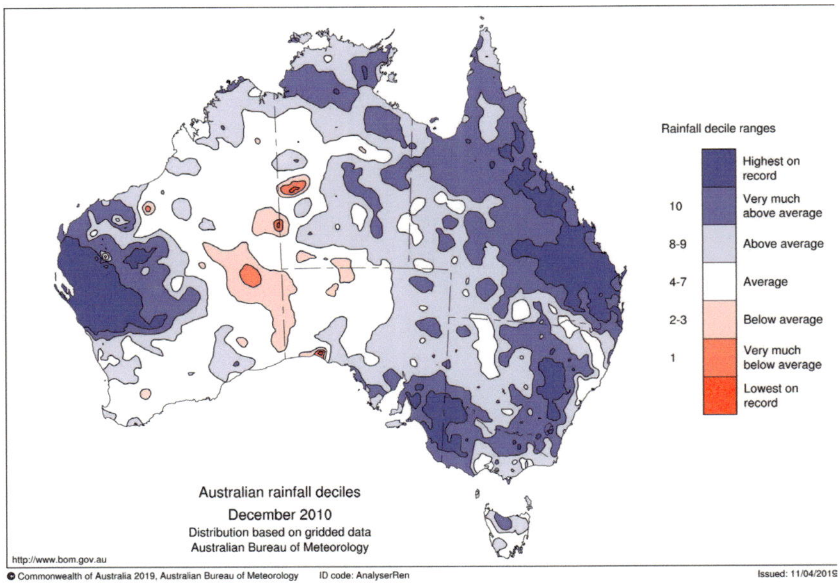

Figure 16.
Australian rainfall deciles for December 2010 derived from the Australian Bureau of Meteorology rain gauge observations.

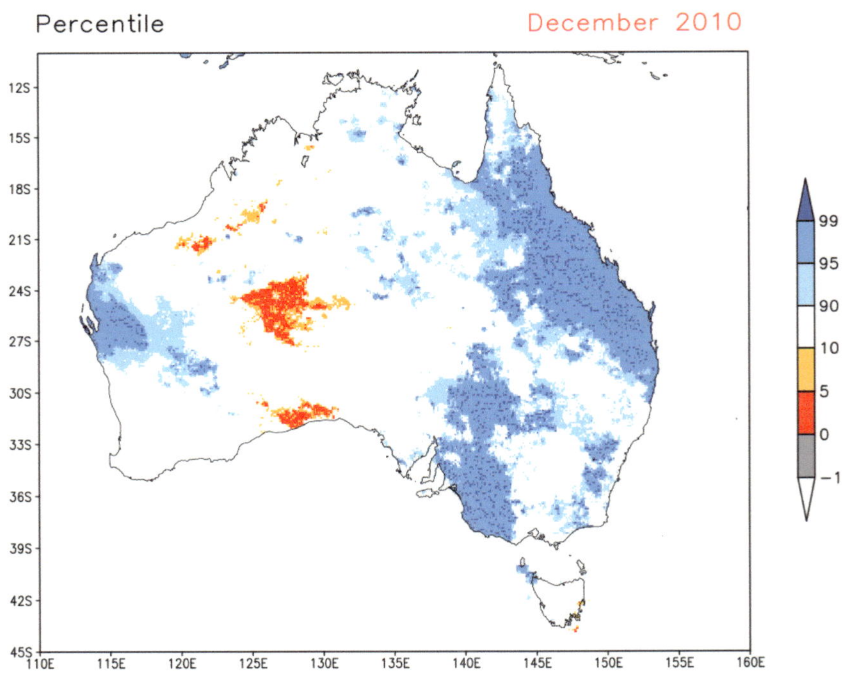

Figure 17.
EORC/JAXA GSMaP rainfall percentile over Australia for December 2010.

the area from 100°E to 105°E; EQ to 8°N is presented in **Figure 19**. In November 2014, accumulated rainfall exceeded 1000 mm along the east coast of Peninsular Malaysia. The first episode of persistent heavy rainfall occurred from 13 to 20 November. In December 2014, areas of monthly total rainfall above 500 mm expand over the southern part of Thailand and the most of Malaysia. Particularly heavy

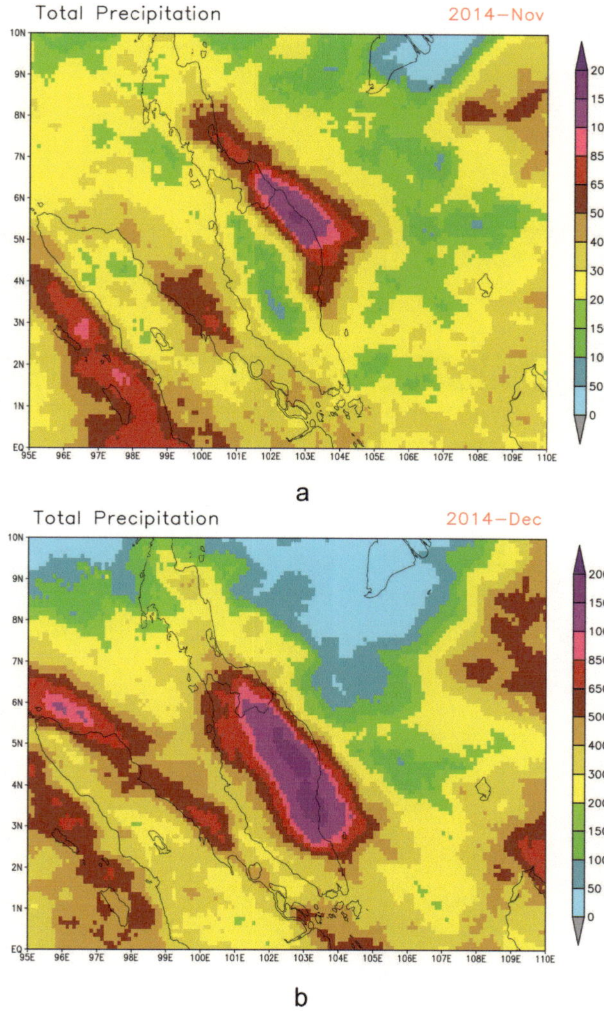

Figure 18.
EORC/JAXA GSMaP total precipitation over Peninsular Malaysia in (a) November 2014 and (b) December 2014.

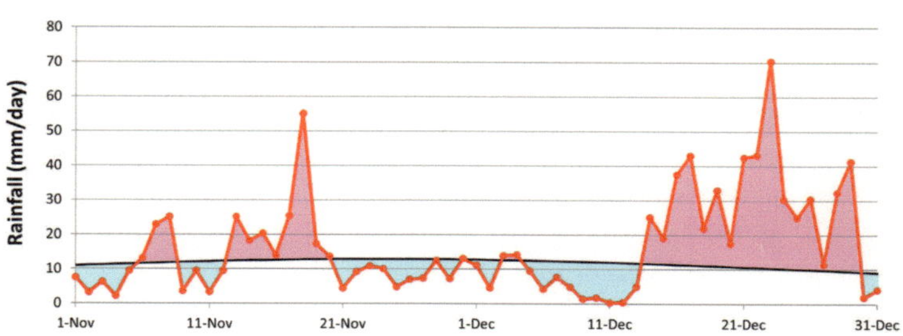

Figure 19.
EORC/JAXA GSMaP time series of daily precipitation for November–December 2014 averaged over land in the area from 100°E to 105°E; EQ to 8°N; solid black line represents 18-year mean.

precipitation occurred over the eastern parts of the Peninsular with monthly total rainfall estimated at 1500 mm and above. Second episode of long-lasting heavy rainfall occurred from 14 to 30 December. Results obtained from space-based observations are in correspondence with results presented in [13].

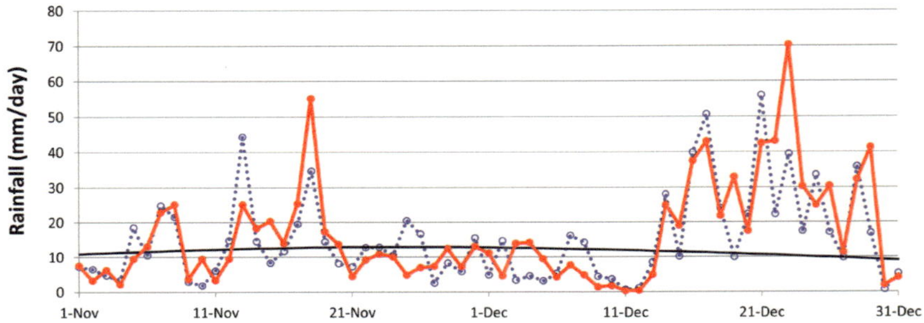

Figure 20.
Time series of daily precipitation for November–December 2014 averaged over land in the area from 100°E to 105°E; EQ to 8°N derived from (a) EORC/JAXA GSMaP (red line) and (b) CPC GAG (blue line).

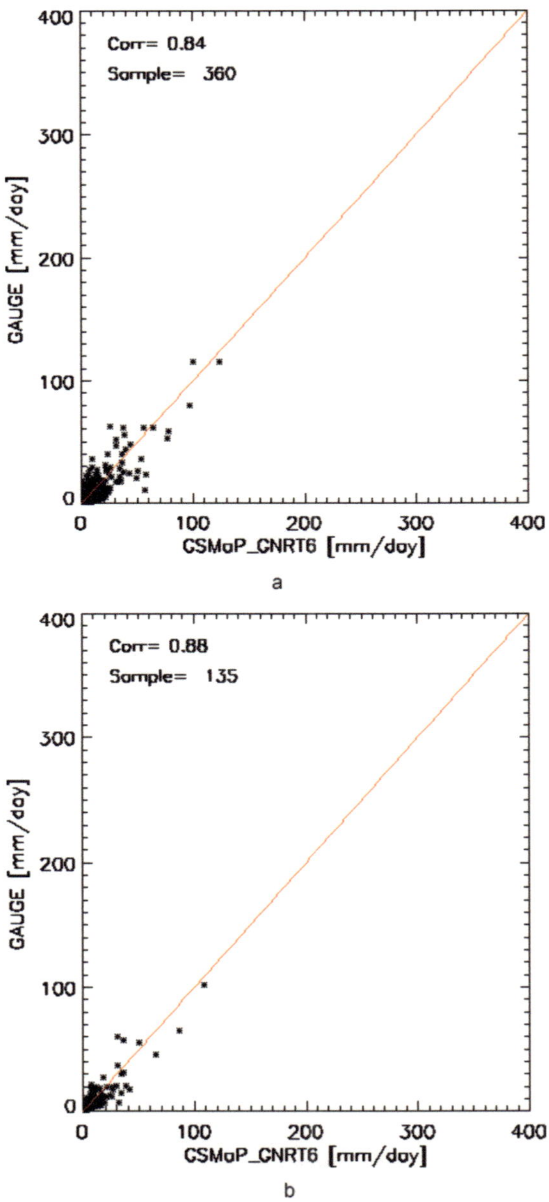

Figure 21.
Scatter plot of (a) pentad (5-day) and (b) 10-day precipitation averaged over land in the area from 100°E to 105°E; EQ to 8°N for November–December 2014 derived from the EORC/JAXA GSMaP versus the CPC GAG.

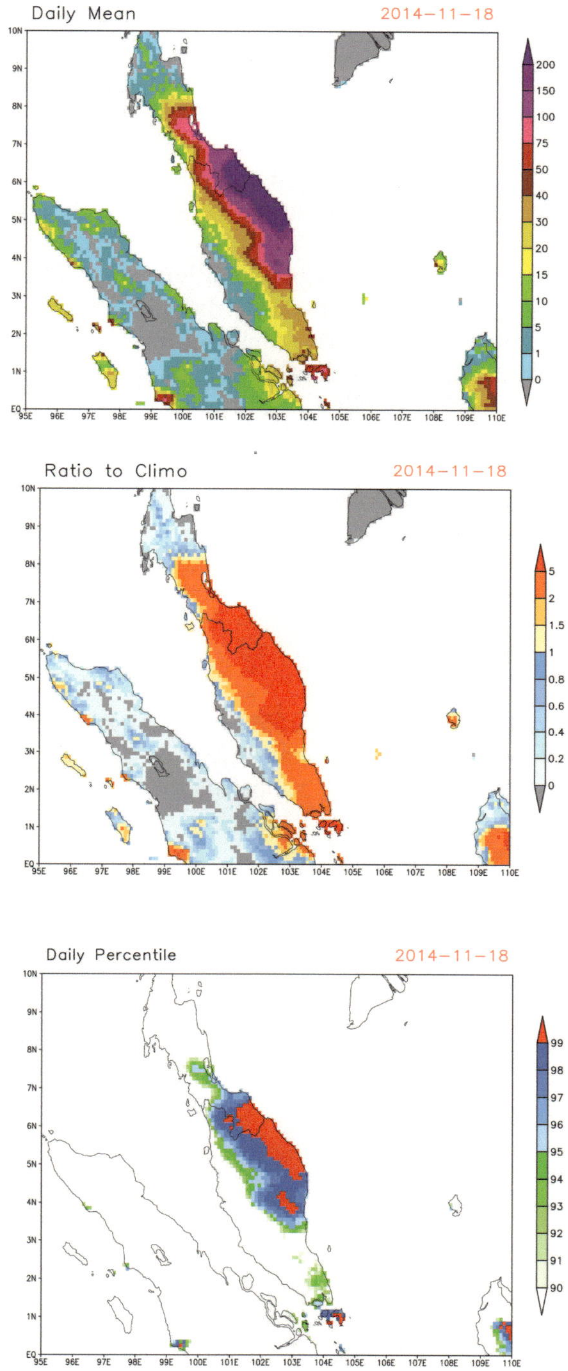

Figure 22.
EORC/JAXA GSMaP (a) daily precipitation over the Peninsular Malaysia on 18 November 2014, (b) ratio of the daily precipitation to 18-year climatology, and (c) daily percentile.

In **Figure 20**, time series of daily precipitation derived from the EORC/JAXA GSMaP data and in situ observations derived from the CPC rain gauge analysis (CPC GAG) for November–December 2014 averaged over land in the area from 100°E to 105°E; EQ to 8°N are presented. All episodes of heavy precipitation in November and December 2014 are detected well by space-based observations. In general, there is a good correspondence between the EORC/JAXA GSMaP space-based rainfall estimates and the CPC GAG rain gauge analysis.

Scatter plot—comparison between CPC GAG and GSMaP satellite precipitation estimates—is presented in **Figure 21**; results for pentad (5-day) and 10-day mean precipitation (mm/day) are plotted on the top and bottom panels, respectively. Only data pairs of precipitation over a 0.25°lat/lon grid box with at least one reporting rain gauge are included in the comparison. The correlation coefficients of pentad and 10 days are 0.84 and 0.88, respectively, indicating good agreement between space-based estimates and surface-based rain gauge observations.

An example of detecting daily heavy precipitation using GSMaP data is presented in **Figure 22**. Precipitation was particularly heavy across the eastern coast of the peninsular with daily totals above 200 mm (**Figure 22a**); this exceeded 18-year climatology more than five times (**Figure 22b**). In Narathiwat province of Thailand and Kelantan and Terengganu provinces of Malaysia, the daily precipitation was higher than the 99th percentile (**Figure 22c**) causing widespread flood in the affected areas.

In summary, presented case studies of detecting extreme precipitation in Australia, Thailand, and the Peninsular Malaysia demonstrate that space-based observations provide valuable information for monitoring heavy rainfall.

5. Conclusions

These first results of implementation of WMO SEMDP demonstrate that space-based estimates of extreme precipitation are an effective solution to enhance capacity of RCCs and NMHSs for monitoring drought and heavy rainfall assisting governments and local communities with informed decision-making in adaptation to climate variability and change.

Acknowledgements

The World Meteorological Organization and the Climate Risk and Early Warning Systems (CREWS) international initiative provided financial support for SEMDP. The EORC/JAXA and the CPC/NOAA developed operational SEMDP extreme precipitation products.

Conflict of interest

Declaration: There is no conflict of interest.

Author details

Yuriy Kuleshov[1,2,3]*, Toshiyuki Kurino[4], Takuji Kubota[5], Tomoko Tashima[5] and Pingping Xie[6]

1 The Australian Bureau of Meteorology, Melbourne, Australia

2 Royal Melbourne Institute of Technology (RMIT) University, Melbourne, Australia

3 The University of Melbourne, Melbourne, Australia

4 World Meteorological Organization, Geneva, Switzerland

5 Earth Observation Research Center, Japan Aerospace Exploration Agency (EORC/JAXA), Tsukuba, Japan

6 Climate Prediction Center, National Oceanic and Atmospheric Administration (CPC/NOAA), Washington, DC, USA

*Address all correspondence to: y.kuleshov@bom.gov.au

References

[1] Stocker TF, Qin D, Plattner G-K, Tignor M, Allen SK, Boschung J, et al., editors. IPCC, 2013: Climate Change 2013: The Physical Science Basis. Contribution of Working Group I to the Fifth Assessment Report of the Intergovernmental Panel on Climate Change. Cambridge, United Kingdom/ New York, NY, USA: Cambridge University Press; 2013. p. 1535. DOI: 10.1017/CBO9781107415324

[2] Asia-Pacific Disaster Report 2012—Reducing Vulnerability and Exposure to Disasters. UNESCAP [Internet]. 2012. Available from: https://www.unescap.org/publications/asia-pacific-disaster-report-2012-reducing-vulnerability-and-exposure-disasters [Accessed: January 13, 2019]

[3] IPCC. AR5 WG I, Annex III Glossary [Internet]. 2013. Available from: https://www.ipcc.ch/site/assets/uploads/2018/02/WG1AR5_AnnexIII_FINAL.pdf [Accessed: January 13, 2019]

[4] Kubota T, Shige S, Hashizume H, Aonashi K, Takahashi N, Seto S, et al. Global precipitation map using satellite-borne microwave radiometers by the GSMaP project: Production and validation. IEEE Transactions on Geoscience and Remote Sensing. 2007;**45**(7, part 2):2259-2275. DOI: 10.1109/TGRS.2007.895337

[5] Xie P, Joyce R, Wu S, Yoo S-H, Yarosh Y, Sun F, et al. Reprocessed, bias-corrected CMORPH global high-resolution precipitation estimates from 1998. Journal of Hydrometeorology. 2017;**18**(6):1617-1641. DOI: 10.1175/JHM-D-16-0168.1

[6] Drought Definition in Australia [Internet]. 2019. Available from: http://www.bom.gov.au/climate/glossary/drought.shtml [Accessed: January 13, 2019]

[7] Annual Australian Climate Statement 2002 [Internet]. 2003. Available from: http://www.bom.gov.au/climate/current/annual/aus/2002/ [Accessed: January 13, 2019]

[8] Annual Australian Climate Statement 2006 [Internet]. 2007. Available from: http://www.bom.gov.au/climate/current/annual/aus/2006/ [Accessed: January 13, 2019]

[9] Annual Australian Climate Statement 2010 [Internet]. 2011. Available from: http://www.bom.gov.au/climate/current/annual/aus/2010/ [Accessed: January 13, 2019]

[10] Annual Australian Climate Statement 2018 [Internet]. 2019. Available from: http://www.bom.gov.au/climate/current/annual/aus/2018/ [Accessed: January 13, 2019]

[11] Kogan FN. Operational space technology for global vegetation assessment. Bulletin of the American Meteorological Society. 2001;**82**(9):1949-1964. DOI: 10.1175/1520-0477(2001)082<1949:OSTFGV>2.3.CO:2

[12] Annual Australian Climate Statement 2011 [Internet]. 2012. Available from: http://www.bom.gov.au/climate/current/annual/aus/2011/ [Accessed: January 13, 2019]

[13] Hai OS, Samah AA, Chenoli SN, Subramaniam K, Mazuki MYA. Extreme rainstorms that caused devastating flooding across the east coast of Peninsular Malaysia during November and December 2014. Weather and Forecasting. 2017;**32**(3):849-872. DOI: 10.1175/WAF-D-16-0160.1

Section 3

Chemical and Physical Processes Associated with Rainfall

Chapter 5

Modeling Aerosol-Cloud-Precipitation Interactions in Mountainous Regions: Challenges in the Representation of Indirect Microphysical Effects with Impacts at Subregional Scales

Ana P. Barros, Prabhakar Shrestha, Steven Chavez and Yajuan Duan

Abstract

In mountainous regions, the nonlinear thermodynamics of orographic land-atmosphere interactions (LATMI) in organizing and maintaining moisture convergence patterns on the one hand, and aerosol-cloud-precipitation interactions (ACPI) in modulating the vertical structure of precipitation and space-time variability of surface precipitation on the other, are difficult to separate unambiguously because the physiochemical characteristics of aerosols themselves exhibit large sub-regional scale variability. In this chapter, ACPI in the Central Himalayas are examined in detail using aerosol observations during JAMEX09 (Joint Aerosol Monsoon Campaign 2009) to specify CCN activation properties for simulations of a premonsoon convective storm using the Weather Research and Forecasting (WRF) version 3.8.1. The focus is on contrasting AIE during episodes of remote pollution run-up from the Indo-Gangetic Plains and when only local aerosols are present in Central Nepal. This study suggests strong coupling between the vertical structure of convection in complex terrain that governs the time-scales and spatial organization of cloud development, CCN activation rates, and cold microphysics (e.g. graupel production is favored by slower activation spectra) that result in large shifts in the spatial distribution of precipitation, precipitation intensity and storm arrival time.

Keywords: aerosol-cloud-precipitation interactions, ACPI, orography, indirect effect, Himalayas

1. Introduction

The aerosol indirect effect (AIE) refers to the cascade of processes (aerosol-cloud-precipitation interactions, ACPI) linking the space-time variability of aerosol physiochemical properties to modification of the vertical structure of precipitation microphysics that result in changes in timing and spatial patterns of precipitation accumulation at the ground. In mountainous regions, orographic modification of atmospheric circulations at multiple scales can further modulate ACPI and consequently have a significant impact on the spatial distribution of precipitation, that is to say the allocation of freshwater input and hydrologic response among adjacent mountain catchments [1–3].

The climatology of the observed strong peak of aerosol optical depth in the premonsoon season in the Indo Gangetic Plains (IGP) has been well documented in numerous studies using satellite retrievals [4–7]. At regional and continental scales, [8] points out that, in the South Asian monsoon region such as the Indian Subcontinent and Himalayas, the net effect of ACPI before and during the monsoon depends on large-scale circulations, moisture availability, and the presence of aerosol hot-spots. Reduction of the efficiency of raindrop dynamics (coalescence and breakup) on account of the presence of very high concentrations of small cloud droplets where aerosol concentrations are very high (aerosol hot-spots) results in delay of precipitation at the surface while very small cloud droplets are transported to higher levels in the troposphere in the direction of storm propagation. Upward transport results in a large population of supercooled drops aloft that freeze, interact with each other to form graupel and hail and subsequently melt as they fall, thus invigorating deep convection through release and production of latent heating at different levels in the troposphere. Higher CCN concentrations from fine aerosol particles slow the conversion of cloud drops into raindrops, thus suppressing rainfall production initially followed by intensification later [9]. Besides the time-delay of precipitation processes, several studies [10–19] have shown that the aerosol effect on cloud microphysical processes strongly depends on specific environmental conditions, varies with cloud types, and thus storm regime. Overall, these studies suggest that mechanisms of aerosol-cloud-rainfall interactions are very complex and highly nonlinear, and therefore transferability and generalization of the results learned from one case study for a particular storm may not be applicable for other storm in different environmental conditions, including climate regime and topography. The latter plays a significant role in airflow modification which in turn strongly impact microphysical pathways. At subregional scales, given similar regional meteorology, these dynamic feedbacks translate into smaller areas of enhanced convective precipitation that is a redistribution of precipitation conditional on aerosol-cloud interactions. Intercomparison modeling studies using CN with different activation characteristics suggest that the timing and intensity of precipitation are tightly linked to regional and subregional scale aerosol characteristics. Recent NWP (Numerical Weather Prediction) simulations in the Southern Appalachians Mountains (complex terrain with moderate elevation <2500 m) show that using regional CCN activation characteristics obtained from field measurements [20] has strong impact on rainfall structure as compared to standard continental aerosol by reducing unrealistic light rainfall on the one hand, and by intensifying convection on the other due to strong modification of cloud microphysics, even more so in the case of local vis-a-vis synoptic forcing [21, 22]. This begs the question of whether the characterization of regional aerosol is not only desirable, but indeed necessary toward achieving a substantial improvement in NWP's predictive skill at high spatial resolution and short time-scales (< 24 hr) toward decreasing phase errors in storm arrival and improving rainfall intensity [2, 23, 24].

Shrestha and Barros [7] identified the central region of the Himalayas and adjacent foothills as a region of potentially high ACPI as the synoptic scale aerosol plume in the IGP penetrates, runs up and accumulates along deep river valleys. Indeed, [25] showed how IGP aerosol can remain sequestered to form pools over low lying areas and valleys in the Middle Himalaya after there is a full retreat of the pollution over the IGP. The aerosol pool is eventually scavenged by the formation of low-level clouds and fog, and washed out by rainfall similar to subregional scale forcing in the inner region of the Southern Appalachians investigated by [21, 22]. Specifically, [26] showed that, in the presence of regional scale aerosol clouds and during dry periods, the mean volume aerosol concentration increased, and so did the aerosol mass concentrations in two different valleys of Central Nepal, the Marshyangdi and the Kathmandu, followed by rain-out. In addition, the topography of the region was found to play an important role in modulating the diurnal cycle of aerosol number concentration due to the diurnal cycle of katabatic and anabatic winds. Previous studies by [27–30] over the central Himalayas in Nepal have shown that the space-time distribution of rainfall and the terrain are strongly intertwined in the region. Depending upon the type of cloud systems and synoptic conditions, changes in aerosol number concentration and chemical properties influence the microphysical pathways of ACPI in different ways, resulting in suppression of rainfall, storm invigoration, and even spatial displacement of rainfall [10, 13, 31–35]. The particle sizes measured during the Joint Aerosol Monsoon Experiment (JAMEX09) in Central Nepal indicate that the dominant aerosol mode is around 100 nm [26], which is also consistent with the predominance of fine aerosol (< 350 nm) found by [36] in the Himalayan foothills using MISR (Multi Imaging Spectro-Radiometer) observations.

The dependence of the aerosol sensitivity on environmental conditions and storm regimes necessitates a better understanding of the joint climatology of aerosol characteristics, regional storm systems and associated precipitation (e.g., premonsoon, monsoon, post-monsoon and winter precipitation in the Himalayas). In-situ measurement of aerosol chemical and physical properties for the different seasons of the year is required to evaluate the sensitivity of the aerosol for different storm regimes using numerical models. Only then, a clear picture of ACPI might emerge. Here, we present an exploratory study to investigate the CCN sensitivity of the numerical simulation of a premonsoon season storm in the Central Himalayas associated with the intrusion of a major IGP aerosol plume (**Figure 1**). The CCN spectra used in the study were estimated from the in-situ measured aerosol size distribution and chemical composition during the Joint Aerosol Monsoon Experiment 2009 (JAMEX09) [26, 37].

The ultimate objective is to investigate ACPI for remote aerosol linked to run-up of a major haze event over the IGP against locally produced aerosol that exhibit very different activation behavior (hygroscopicity) even when concentration numbers are not significantly different [26, 37]. Because this study consists of simulations of the same storm system using different CCN, it allows us also to assess quantitatively the likely impact of changes in storm dynamics on precipitation fields at the ridge-valley scale in the Middle Himalaya caused by IGP pollution. Significant shifts in the maxima of the event cumulative precipitation were first observed between the simulations conducted with control continental aerosol spectra in WRF and from JAMEX09 by [38]. Differences in the simulated vertical profile of temperature, water vapor mixing ratio and hydrometeor distributions (indicative of differences in latent heat absorption/release) lead to changes in local circulations, which in turn are tied to landform. In particular, this study suggests strong coupling among CCN activation spectra, the vertical structure of convection in complex terrain, and cold microphysics (e.g., graupel formation) that strongly impacts the spatial distribution of precipitation at the surface. Finally, we discuss the results in the context of regional hydrometeorology and impact on spatial patterns of precipitation accumulation that result from changes in space-time storm evolution displacing convective cells among adjacent catchments in Central Nepal including the Kulekhani Water Reserve (KWR) hydropower dam, which provides critical electricity to Kathmandu, and the Indrawati basin (IDR), the headwaters of the Sun Koshi river that links central to eastern Nepal (**Figure 2**). The chapter is structured as follows: Section 2 describes the experimental setup for the simulation. Results are discussed in Section 3. Summary and conclusions are presented in Section 4.

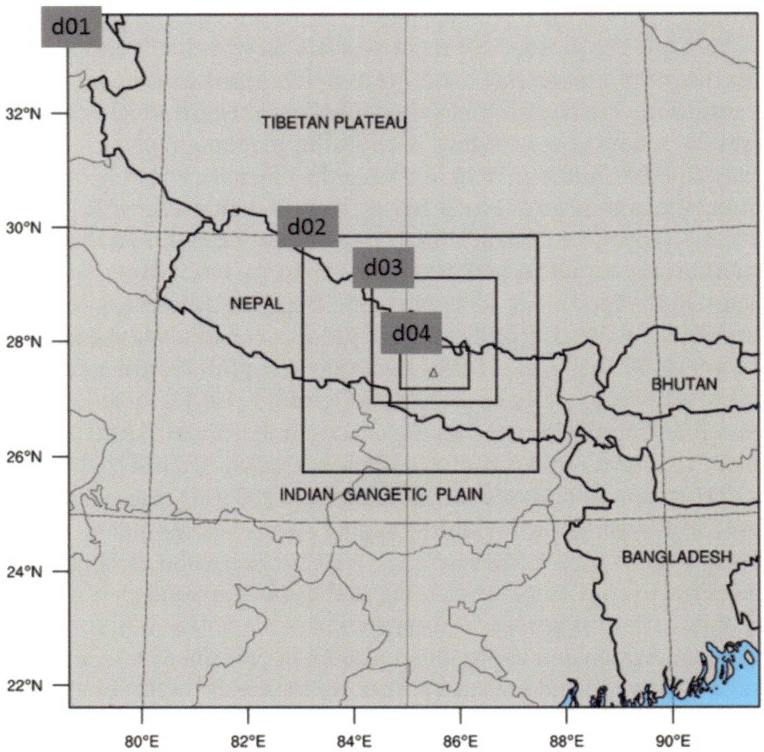

Figure 1.
Map of the region of study. The four nested domains used for WRF model simulations are centered over Central Nepal. Domain d01 (27 km) encompass the Indian Gangetic plain (IGP) and the Tibetan plateau extending up to Bhutan and Bangladesh in the east. Domains d02 (9 km) and d03 (3 km) are over Central Nepal. Domain d04 (1 km) encloses the Kathmandu Valley (marked with triangle).

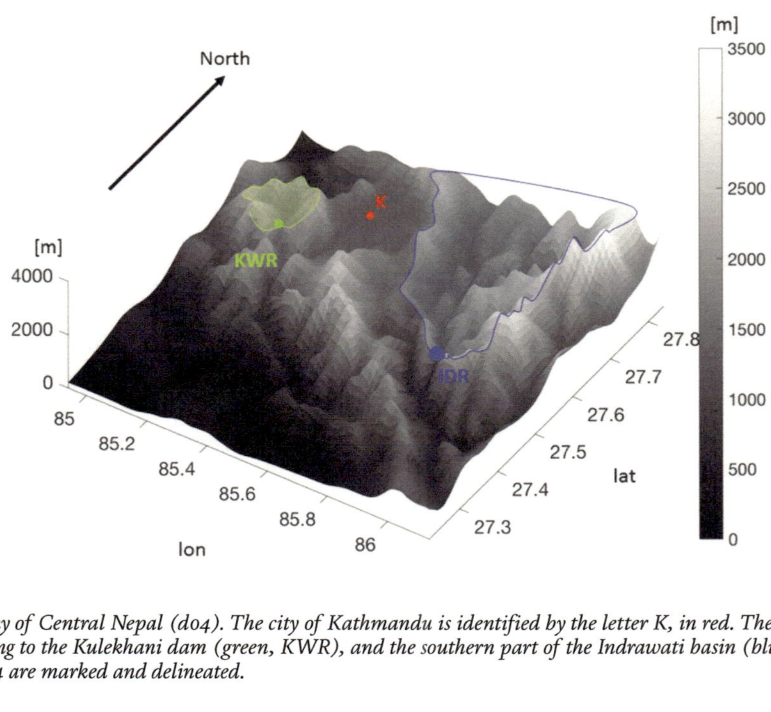

Figure 2.
Topography of Central Nepal (d04). The city of Kathmandu is identified by the letter K, in red. The catchment contributing to the Kulekhani dam (green, KWR), and the southern part of the Indrawati basin (blue, IDR) within d04 are marked and delineated.

2. Numerical experiments of ACPI sensitivity to CCN

2.1 WRF model setup

The Advanced Weather Research and Forecasting (WRF) model Version 3.8.1 [39] was used for numerical simulations of a northwesterly convective storm over Central Nepal on May 15–16, 2009 during JAMEX09. The model configuration was set up similar to [3] with four one-way nested domains with horizontal grid spacing of 27-, 9-, 3-, and 1-km, corresponding to grid sizes of 51 × 51, 52 × 52, 73 × 73, and 121 × 73 for the first (d01), second (d02), third (d03), and fourth (d04) domains, respectively (**Figure 1**). In order to resolve low-level cloud formation and precipitation processes, a terrain-following vertical grid with 90 layers was constructed with 30 levels in the lowest 1 km AGL and the model top at 50 hPa. WRF simulations during a two-day period were conducted starting at 00:00 UTC 14 May 2009 (5:45 LT in Nepal) and ending at 00:00 UTC May 16, 2009. The first six hours of simulation were disregarded for analysis.

Initialization and lateral boundary conditions are updated every 6-hours and interpolated in-between using the National Centers for Environmental Prediction (NCEP) Final Operational Global Analysis (FNL) with 1 × 1° horizontal resolution. The Kain-Fritsch cumulus parameterization scheme [40] is used in the first and second domains (27 and 9 km resolution), and convection is resolved explicitly in the third (3 km) and fourth (1 km) domains. Other physics options include Milbrandt and Yau's 2005 (MY05) double moment microphysics [41], a new version of the Rapid Radiative Transfer Model radiation scheme for longwave and shortwave [42], and the unified Noah land-surface model [43] applied for all four domains. Following [2, 3], the Mellor-Yamada-Nakanishi-Niino (MYNN) planetary boundary layer scheme [44] is selected along with the Monin-Obukhov (Janjic Eta) surface layer scheme to better capture low level cloud formation. The soil temperature and moisture fields are also initialized from the NCEP FNL data.

2.2 Modeling experiments with Milbrandt-Yau microphysics

The MY05 double moment microphysics scheme (total number concentration and mixing ratio) is used here to investigate the effects of aerosol properties on the sensitivity of ACPI. Number concentrations of nucleated cloud droplets (N_{CCN}) in MY05 are calculated based on a four-parameter CCN activation spectrum proposed by [45], hereafter referred to as CBP98. This CCN activation scheme has demonstrated improved estimation of cloud droplet numbers as it accounts for the depletion of small-sized condensation nuclei (CN) with increasing supersaturation:

$$N_{CCN}(S_{v,w_{max}}) = C S_{v,w_{max}}^{k} F\left(\mu, \frac{k}{2}, \frac{k}{2} + 1; -\beta S_{v,w_{max}}^{k}\right) \quad (1)$$

Where $s_{v,w_{max}}$ is the maximum water vapor supersaturation and F(a, b, c; x) is a hypergeometric function. The four fitted parameters in Eq. (1) can be interpreted as follows: C is a scaling factor, *k* is the slope of the linear relationship between log of N_{CCN} and log of s_v, and β indicates the location of the slope break between the fast (linear) CCN activation regime governed by *k* at lower supersaturation and the slow regime described by the shape parameter μ at high supersaturation (see **Figure 1** in CPB98).

CPB98 [45] fitted two CCN activation spectra respectively for "representative" maritime (CCN1, Type 1) and for continental (CCN2, Type 2) aerosol which are available in the standard MY05 parameterization in WRF, but the formula in Eq. (1) and corresponding fitting parameters for each aerosol type are not directly employed for computing N_{CCN} in the microphysics scheme. Instead, maximum supersaturation is first expressed as a function of updraft speed w, temperature T, and pressure p using an iterative method outlined by [46]. To reduce computational costs, non-linear least-square fits are applied subsequently to $s_{v,w_{max}} = f(w, T, p)$ and $N_{CCN} = f(w, T, p)$ for the specified CCN spectra. For the present study, two additional CCN spectra derived from ground-based observations at Dhulikhel [26, 37] in the Kathmandu valley during JAMEX09 were incorporated into MY05 using non-linear regression fits as described in [38, 41]: (1) CCN3 (May 15, 2009) corresponding to conditions during a large-scale haze event in the Indo-Gangetic Plains (IGP) with aerosol run-up in the Central Himalayas as described by [25], thus remote aerosol; and, (2) CCN4 (May 16, 2009) corresponding to locally produced aerosols after washout of remote aerosol by rainfall in the previous day, thus local aerosol.

The CCN spectra collected before the rainfall (CCN3) and the day after (CCN4), are shown along with the standard continental (CCN2) and marine (CCN1) types from CPB98 [45] in **Figure 3**. Compared to the continental CCN2 type, CCN3 shows lower CCN number concentrations for supersaturation $S_c < 0.1\%$ and higher CCN number concentrations at higher supersaturation. The N_{CCN} for CCN4 is always lower than CCN3, but it approaches and even exceeds CCN2 for supersaturations >0.3%. This is attributed to high number concentrations of small aerosol particles that are activated at high supersaturation. The marine spectrum CCN1 (displayed only for contrast against the continental aerosol spectra) shows CN depletion halting activation at low values of supersaturation close to 0.1%.

Here, three WRF 3.8.1 simulations of regional weather on May 14–15 2009 are used to probe the sensitivity of microphysical and dynamical processes to the aerosol indirect effects (AIE) including: a control run (hereafter, CCN2) using the

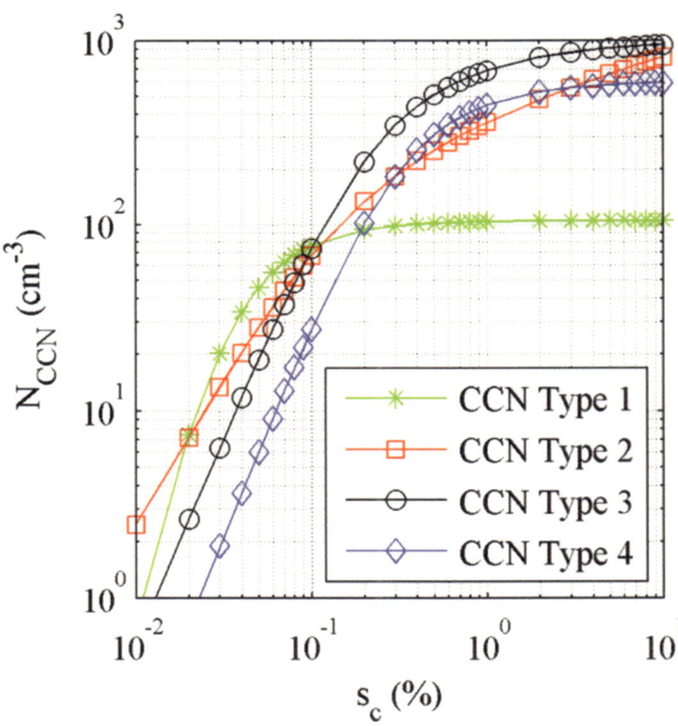

Figure 3.
The marine (CCN type 1) and continental (CCN type 2) CCN spectra are based on the study by [45]. The CCN spectra for May 15 (CCN type 3) and May 16 (CCN type 4) are the estimated spectra from aerosol size distribution and chemical properties at Dulikhel in Central Nepal during JAMEX09 campaign [26, 37, 38]. CCN3 represents remote aerosol intrusion from the IGP. CCN4 represents local aerosol sources.

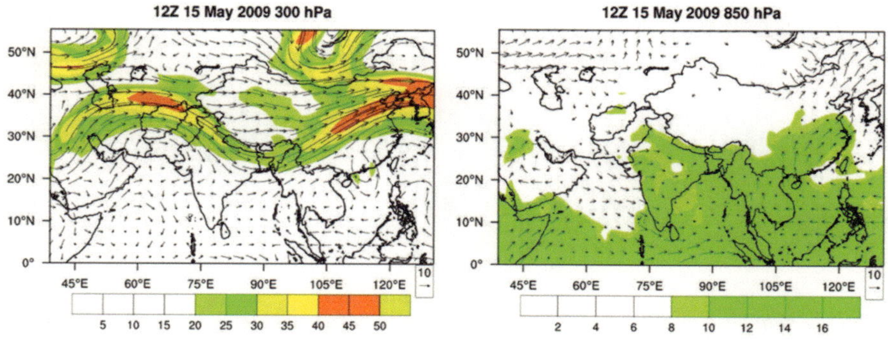

Figure 4.
ERA interim reanalysis. Left: Wind vectors and wind speed contours at 300 hPa showing the upper level westerly jet in the free troposphere above the terrain. Right: 850 hPa wind field and mixing ratio, showing the influx of moisture to Central Nepal and the IGP where the low level flow is blocked by the terrain. https://www.ecmwf.int/en/forecasts/datasets/archive-datasets/reanalysis-datasets.

default continental aerosol in the MY05 microphysics scheme from CBP98, and two simulations, CCN3 and CCN4, with JAMEX09 aerosol spectra from [26, 37, 38]. The differences in model configuration among the three simulations are limited to the cloud nucleation scheme in MY05. Direct radiative effects of aerosol and the parameterization of nucleation efficiency of different ice nuclei on heterogeneous drop freezing, which could be important in ice-phase clouds [47, 48] cannot be addressed in the WRF configuration used in this study, but it is possible to examine AIE in contrasting large-scale environments, specifically during the propagation of a convective afternoon storm on May 15, second day of simulation.

The synoptic scale conditions on May 15 were characterized by northwesterly flow aloft with the axis of the upper level trough, east of the study region (**Figure 4**). The NW flow extends up to 3 km amsl. At lower levels, a cyclonic circulation is present in the IGP, creating a south-easterly flow upon the outermost foothills of the Himalayas over central Nepal. This low level flow also tends to veer along the hills westward and turn clockwise, impinging the outer ridges of Kathmandu Valley and the KWR from north-west, in agreement with the direction of upper level flow. In due course, this potentially unstable air in the lower troposphere releases its instability as it flows above the topography, triggering cells of intense precipitation. Previous studies of orographic effects on rainfall in Mesoscale Alpine Program (MAP) have also pointed out the importance of the change in static stability of the flow at low levels coupled with the orographic modification of the flow in the prediction of the intensity, location and duration of orographic precipitation [49].

3. Results and discussion

The results are analyzed with a focus on identifying and explaining differences and similarities among precipitation accumulation patterns and among microphysical vertical structure for the three different CCN activation spectra (CCN2, CCN3 and CCN4). Rain gauges operated by the Nepal Department of Hydrology and Meteorology (DHM) mostly registered zero precipitation during the 2 days of study, except in the northwest sector of d04 (**Figure 1**) were observed precipitation totals are on the order of 5–10 mm and up to 60 mm consistent with model simulations [38]. **Figure 5** shows the total precipitation accumulations for the 2nd-day of simulations draped over the 3D topography looking from the High Himalaya (north) to the IGP (south). Note the organization of rainfall hot-spots along the

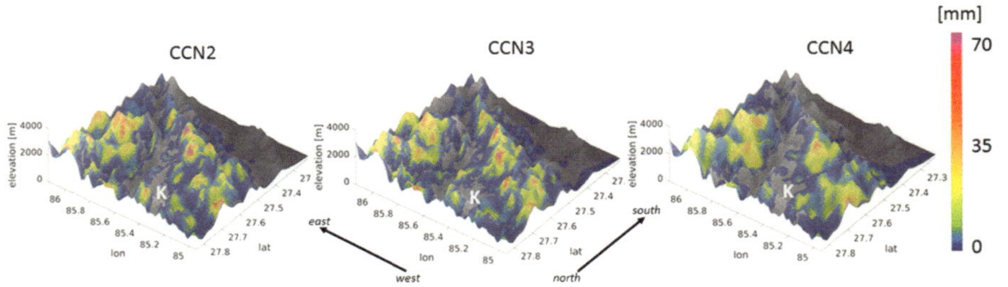

Figure 5.
*Simulated precipitation accumulation patterns on 15 May, 2009 in d04 (**Figure 1**) at 1 km spatial resolution. K indicates the location of Kathmandu.*

ridges to the west and east of Kathmandu valley, and the much smoother patterns with more widespread low rainfall accumulations for CCN4.

Radar measurements of precipitation are non-existent in this region, and thus TRMM (Tropical Rainfall Measurement Mission) and GPM (Global Precipitation Measurement) since 2014, as well as Terra and AQUA satellite overpasses provide valuable spatial data of respectively rainfall and cloud-coverage over the region of study [7, 25, 50]. Nevertheless, the satellite trajectories vary substantially from one overpass to the next, and the observations may have therefore a limited use for storm-based case-studies depending on timing and geography. In particular, the paths of the A-Train Satellites are east and west of the innermost d04 domain on May 15, 2009. By contrast, geostationary satellites such as Meteosat7, Fengyun-2D, and Kalpana provide a regional overview of cloud processes in this region. A survey of the Meteosat7 imagery shown by [38] indicates that the storm strengthens after 0900 UTC organizing itself along the outer ridges of the Kathmandu Valley consistent with the higher precipitation accumulations on the western ridges in **Figure 5**. As described earlier, the north-westerly storm is fueled at low levels by south-easterly flow impinging on the southern and western ridges of the Kathmandu valley producing a series of multiple convective cores, enhanced by the complex orography (**Figure 4**). Consequently, the cumulative WRF rainfall shows strongly localized maxima aligned with topography, especially in the case of CCN2 and CCN3, whereas CCN4 fields are smoother with lower maxima over the western ridges, which becomes suppressed toward the foothills and the IGP.

The differences in spatial accumulation patterns (**Figure 6**, left panels) and cross-section (right panels) between CCN2 (dotted line) and CCN3 (red) are much smaller than the differences between CCN3 and CCN4 (blue). CCN2 and CCN3 closely follow the terrain with heavier precipitation at high elevations including isolated peaks of very high rainfall in the north and west sectors of study domain consistent with the direction of approach of the convective storm. The differences between CCN3 and CCN4 are very large to the south and west especially, with CCN4 producing higher rainfall only to the east.

Overall, the simulated rainfall patterns over the ridges surrounding Kathmandu Valley and the dry patterns in the valley itself are consistent with the patterns of observed rainfall where gauges exist. CCN2 precipitation is higher than CCN3, and both CCN2 and CCN3 are much higher than CCN4. The spatial patterns of [CCN2-CCN3] and [CCN3-CCN4] in **Figure 6** exhibit the same overall spatial patterns of negative and positive differences distributed over the same subregions. Interestingly, both CCN2 and CCN3, yield lower rainfall amounts over the KWR and the IDR catchments. Whereas it is not possible to extrapolate based on one single storm, this result illustrates how the two CCN cases representative of remote aerosol intrusion would produce lower rainfall amounts over two critical landmarks for hydropower and water supply in the region. To extract statistically

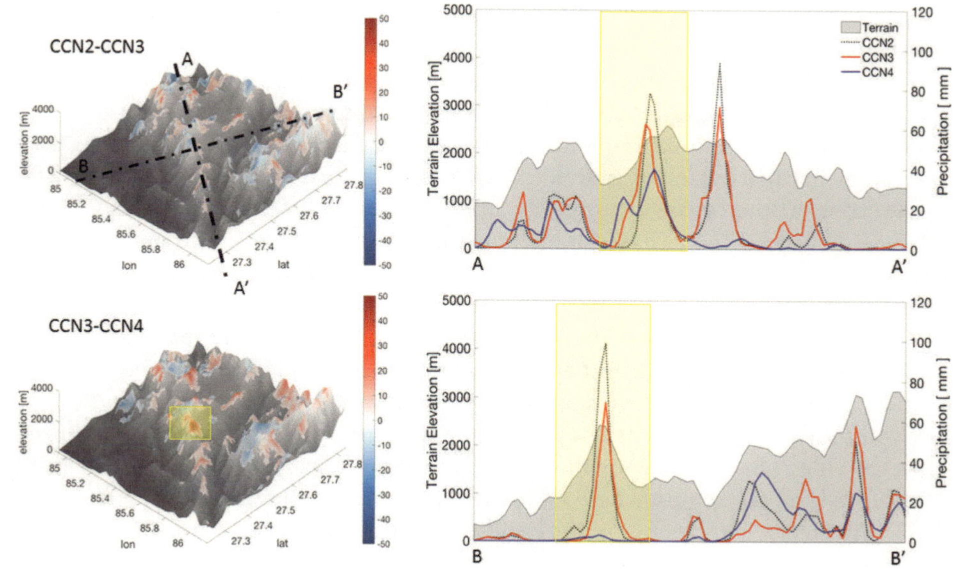

Figure 6.
*Left: Differences in spatial accumulation patterns on May 15, 2009 wrapped on the topography. Right: Cumulative rainfall along cross-sections AA' (north-south) and BB' (west-east) marked on top right panel. The topography along the cross-sections is marked in gray. The yellow box denotes the subdomain used to integrate precipitation in shown **Figure 7**.*

robust information at climate time-scales, it would be necessary to produce simulations for a large number of representative storms, which is out of the scope of the present manuscript.

In the western ridges, the temporal evolution of rainfall (**Figure 7**) over the region defined by two rainfall peaks in CCN2 and CCN3 (**Figure 6**, yellow box) shows that CCN4 (local aerosol) rainfall is late compared to CNN2 and CCN3 (remote aerosol). Further, even if the second peak on the cross-section AA' (**Figure 6**, top right panel) is missed, the rainfall distribution is more filled in the case of CCN4 and when it is raining there is no significant difference in rainfall

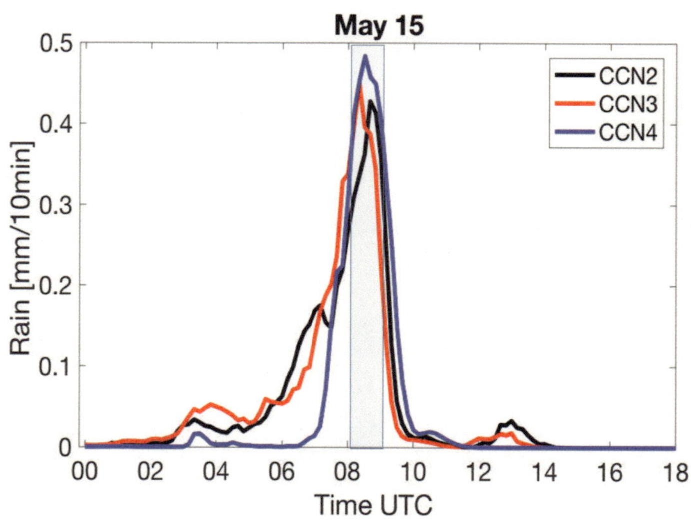

Figure 7.
*Time-series of precipitation integrated within the subdomain define by the yellow box in **Figure 6**. The light blue bar highlights the period used for hydrometeor analysis in **Figures 10** and **11**.*

intensity. The delay of rainfall in CCN4 can be in part attributed to the slopes of the activation spectra shown in **Figure 3**.

Given sufficient moisture convergence at low levels, the steeper slopes of CCN2 and CCN3 lead to more rapid cloud development where orographic updrafts develop (**Figure 8**), and consequently earlier rainfall. In addition, the ACPI feedback on the intensification of individual convective cells is apparent from contrasting the strength and organization of the updrafts from 8:00 to 9:00 UTC between CCN2, CCN3 and CCN4. That is, higher rainfall accumulations at the surface are co-organized with stronger vertical updrafts and downdrafts, and thus faster hydrometeor turn-over times. However, in the case of CCN4 the updrafts are tilted in the north-westerly direction of storm propagation (e.g., 8:20 UTC in **Figure 8**) with development of a wind gust front at low levels, which favors longer-lasting convective cells in the presence of favorable wind shear, specifically northwesterly flow aloft and southeasterly flow at low levels. The CC' cross-section at lower elevations below the ridge (AA') on the upwind slopes of the western ridges just south of KWR allows us to look at a region where CCN4 also produces higher rainfall amounts than CCN3 (**Figure 9**). The strong vertical updrafts in the green region in CCN4 correspond to convective cells forming on the steep slopes ahead of topographic peaks as low level westerly flow (**Figure 4**) from the IGP is orographically forced in the mid-Himalayas (**Figure 2**) and maintained by moist air convergence enhanced by the passage of the storm from 8:00 UTC past 9:00 UTC.

The CCN4 updrafts remain locked in space but strongly tilted with height due the strong northwesterly flow aloft (**Figure 9**, 8:20 UTC). Note the differences between CCN3 and CCN4 along CC': at high elevations, convective CCN3 cells that form in the area defined by the green bar are very short-lived (dissipate before 8:30 UTC) producing lower rainfall amounts, whereas persistent weak and shallow updraft cells locked to the topography at lower elevations in the foothills in CCN3 and in CCN4 produce little precipitation (outside of green region of interest). Interestingly, in this case the vertical structure of CCN3 and CCN4 winds is very similar indicating largely inactive moist processes, and thus no significant impact from CCN activation differences. To further examine the microphysical impacts of the three CCN types on the space–time evolution of the rainfall event, a detailed analysis of the water budget (rainwater, liquid water and ice water) is conducted next.

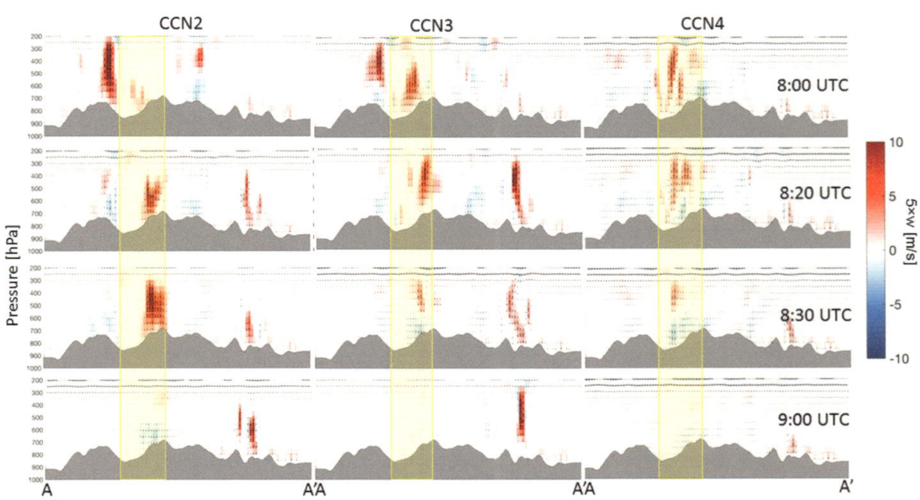

Figure 8.
*Vertical distribution of westerly winds (u) and vertical winds (w) along cross-section AA' during the time of peak storm activity in **Figure 7**. As note in the legend, the vertical wind velocities were multiplied by five to stretch the scale. The yellow stripe marks approximately the location of yellow sector marked in the **Figure 6**, top right panel.*

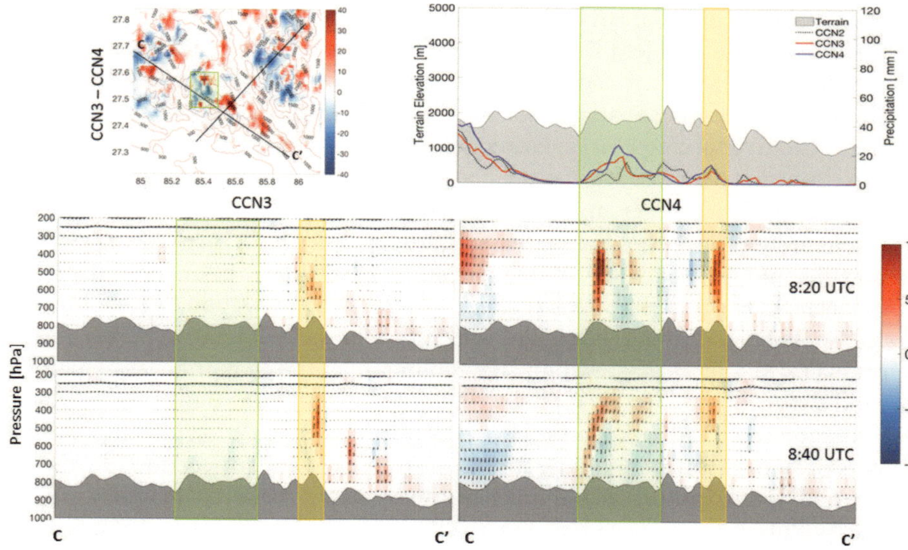

Figure 9.
*Top left: Difference between CCN3 and CCN4 cumulative rainfall on May 15, 2009. Cross-section CC' is downslope of the ridge (AA') on the upwind side at low levels. BB' from **Figure 6** also shown for reference. The green box denotes the subdomain used for analysis. Top right: Cumulative rainfall along cross-section CC' (north–south). Bottom rows: Vertical distribution of westerly (u) and vertical (w) winds along CC'. Orange bar marks location for temporal analysis in **Figure 13**.*

Figure 10 shows the instantaneous spatial fields of Liquid Water Path (LWP, bottom row) and Ice Water Path (IWP, top row) for d04 at 8:30 UTC. Whereas some temporal and spatial variability is expected, note the consistently larger values of IWP from left to right which are representative of the overall behavior for each simulation. The cells of heavy surface rainfall, high LWP and high IWP are spatially coupled to the strong updrafts in CCN2 and CCN3. However, in the case of CCN4, even at 8:30 UTC which is the time of peak rainfall (**Figure 7**) in the subregion of interest, there is a gap between the position of LWP maxima (along CC') and the IWP maxima (along AA"). This is consistent with orographic warm rain processes tied to strong orographic updrafts tilted with height upstream of the ridgeline AA' (not shown), whereas at higher elevations, cold microphysics become dominant.

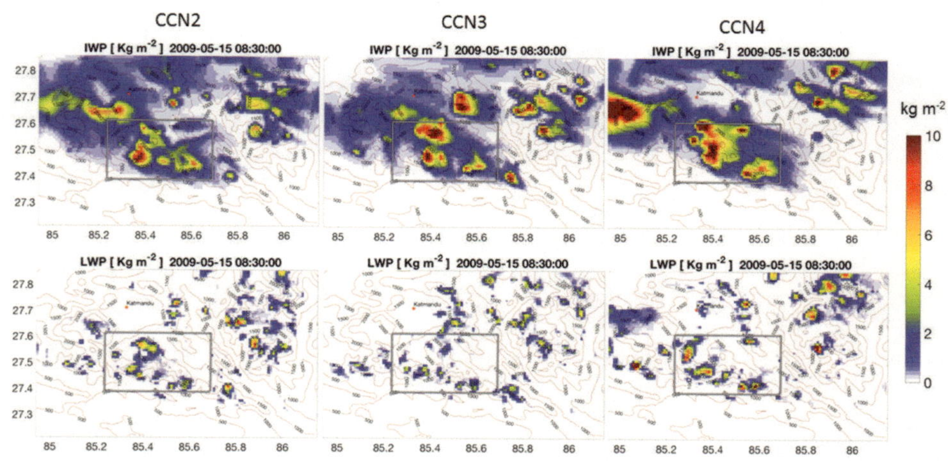

Figure 10.
*Ice water path (IWP, top row) and liquid water path (LWP, bottom row) at 8:30 UTC. The gray box delimits the area used for analysis of the vertical distribution of hydrometeors in **Figure 10**.*

The horizontal extent of the anvil of deep clouds is expected to be larger in the presence of an upper level jet [51], and this was also the case in this simulation as the strong upper level northwesterly winds aligns the anvils of the clouds along AA" in **Figure 10**. The different spatial extent of the anvils of the convective cells also influences radiative forcing (aerosol direct effect), and consequently the energy budget in the atmosphere and at the surface further complicating aerosol-cloud interactions, but this is not examined in this study.

Previous studies have shown that when there are significant increases in aerosol concentration and high numbers of cloud droplets, the efficiency of the conversion of cloud droplets to rainfall decreases along with increasing evaporative cooling of cloud droplets, resulting in increased intensity of downdrafts, cold pools, and gust fronts [13, 35]. Increased evaporative cooling due to large numbers of small droplets also affects the glaciation of shallow clouds (cloud top <5 km) by reducing autoconversion rates and secondary ice crystal production [52]. Smaller and weaker cold pools provide less forcing to the formation of secondary convection, decreasing the overall intensity and spatial reach of the convective storm. [19] found that simulated polluted storms with higher aerosol concentrations produced smaller and weaker cold pools, where the polluted aerosols were defined in terms of higher aerosol concentration. Further, [53] showed that in the absence of graupel in the representation of ice processes in the model microphysics scheme, low level downdrafts are weakened, and the simulated storm moves slower, with more rain and ice converted to snow. Snow can be transported longer distances due to their low density, leading to opposite sign impacts on precipitation due to changes in CCN concentration for simulations with and without graupel.

The spatial and temporal averages of the vertical distribution of hydrometeors for each simulation between 8:00 and 9:00 UTC are shown in **Figure 11**. The number concentrations of the various hydrometeors are not significantly different except for the cloud droplet numbers that are one order of magnitude larger in CCN4. CCN4 mixing ratios are higher for all hydrometeors. Hail forms with very

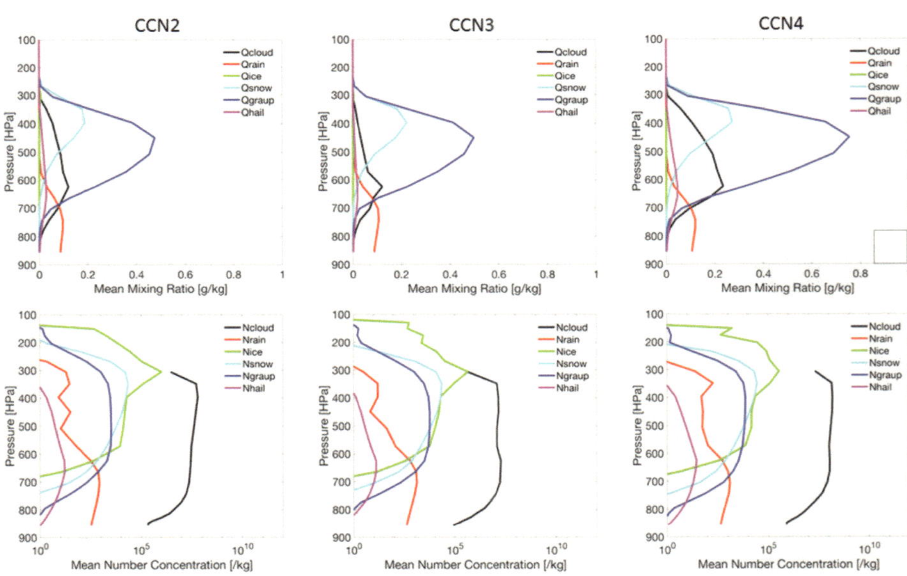

Figure 11.
Mean vertical distribution of hydrometeors within the domain defined by gray box in ***Figure 9*** *between 8:00 and 9:00 UTC on May 15, 2009. Top row: Mean mixing ratio of mixing ratio (top row) and mean number concentration (bottom row). QXXX- mixing ratio, where XXX: Cloud liquid water, rainfall, cloud ice; snow; graupel; and hail. NXXX- number concentration.*

low concentrations and mixing ratios (pink lines) above 700 hPa, the cloud mixing ratios are high well above the freezing level especially for CCN4, thus facilitating interaction of supercooled cloud droplets with cloud ice to form snow aloft, and riming of snow hydrometeors to form graupel between 650 and 400 hPa. This is also apparent by inspecting the changes of the vertical mean number concentration of raindrops and hail on the one hand, and the decreases in the numbers of snow, whereas the mean number concentration of graupel and cloud droplets changes in parallel down to 600 hPa. The graupel mixing ratio decreases at lower levels as it settles fast and eventually melts adding to rainfall below 700 hPa. Because of the flow separation in CCN4 with weak downdrafts at low levels and weak updrafts at high levels over the ridge, the residence times of hydrometeors above 550 hPa are much longer for CCN4 which explains the higher mixing ratios of CCN4 clouds, the much higher mixing ratio of graupel, and hence the large anvils captured by the IWP fields in **Figure 10** (top right panel).

Microphysics budgets and tracking of dynamical feedbacks can be more easily carried out in simulations of isolated storm events [54] or under idealized and controlled conditions [33]. The present study consists of a succession of multiple cells and airflow over complex topography, and thus the approach is to conduct the analysis over selected subdomains within d04. Moreover, the sensitivity analysis is strictly focused on the differences in aerosol activation behavior, not on number concentration. The number of activated CCN2 will be higher than CCN3, and CCN3 will be higher than CCN4 as long as the maximum supersaturation is below 0.3% (**Figure 3**). That is, aerosol activation rates are slower for local aerosol (CCN4) than remote aerosol (CCN3) over a significant range of supersaturation values critical for cloud development, and therefore the initial numbers of hydrometeors are lower for CCN4, which delays precipitation (e.g., **Figure 7**) and forces secondary convection that is enhanced by orographic forcing on the upwind slopes (**Figure 9**) to maintain long-lived strong updrafts that produce deeper clouds with higher mixing ratios (**Figures 10** and **11**).

Although it would be expected that the horizontal advection of snow hydrometeors (low density) aloft should be important due to the presence of strong upper level northwesterly winds, the averaged vertical profiles simulations show a deep system (cloud tops >10 km) in which the dominant ice hydrometer is graupel with peak mixing ratios of 0.45 g/kg for CCN2 and CCN3 and 0.65 g/Kg for CCN4 averaged over the subdomain marked in **Figure 10** and over one-hour period (**Figure 11**). At specific times and locations, the average peak values are as high as 1.5 g/kg for CCN4 vis-a-vis 0.5 g/kg for CCN3 as shown in **Figure 12**.

Ref. [54] reported similar mean vertical profiles for cloud ice and snow in the IOP2A case-study during the Mesoscale Alpine Programme (MAP). The convective system of IOP2A also had a vertical extent exceeding 10 km, and large

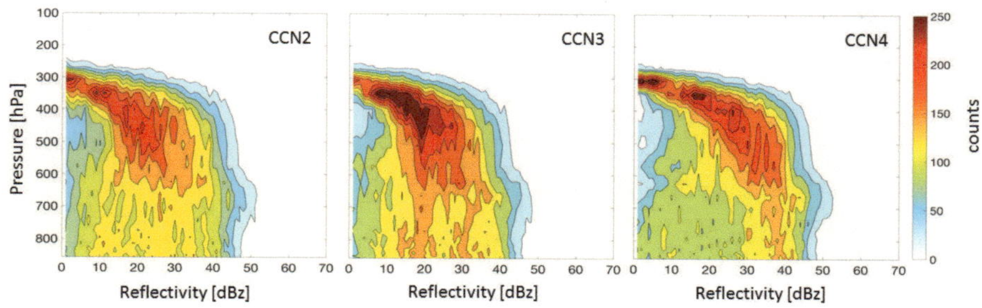

Figure 12.
Contour frequency by altitude diagrams (CFADs) of reflectivity within the gray box domain in ***Figure 10*** *between 8:00 and 9:00 UTC on May 15, 2009.*

amounts of graupel and hail above the freezing level albeit with much lower maximum mean mixing ratio for graupel of 0.15 g/kg and higher for hail at 0.1 g/kg. In this study, the mean hail mixing ratio is less than 0.05 g/kg. Strong and sustained uplifting of low-level moist air brings enough cloud water (with inhibited coalescence to produce rainfall, and thus high number concentrations of cloud droplets) above the freezing level to favor graupel production. This is confirmed by the significantly larger concentration of graupel in CCN4 compared to CCN2 and CCN3 (**Figures 11** and **12**). Note that the mean maximum mixing ratio of cloud water was 0.15 g/kg for IOP2A (MAP) close to the values in this study in **Figure 11** but significantly lower than instantaneous values at local places (**Figure 12**).

Figure 13 shows the Contoured Frequency Altitude Diagram (CFAD) of reflectivity for CCN2, CCN3 and CCN4 corresponding to **Figure 11** as per [55]. The probability contours are expressed in number density. The CFAD is a synthesis that overcomes the mismatch in space and time of simulated precipitation using different CCN initializations, and thus enables examining the changes in ensemble properties of the storm among different simulations. In the early stages of the storm, the distribution is broad and homogeneous at all levels (not shown here). During the mature stages of the storm, the fallout of larger hydrometeors (e.g., large raindrops and melted graupel) causes the distribution to narrow especially below the melting level (see **Figure 12**). Deep convective activity can be identified as the core region with reflectivity higher than 40 dBZ [56]. The CFAD convective core appears to be stronger with a narrow distribution in the CCN4 compared to CNN3 and CCN2, which is consistent with the budget analysis and the mean hydrometeor profiles in **Figure 11**.

It is possible that the model simulations presented here produce excessive graupel, a concern that has arisen in some studies in the past when model simulated reflectivity is compared to radar observations [57, 58]. To address these concerns, the microphysical parameterizations can be calibrated to adjust the relative proportion of graupel versus snow by for example lowering overall riming efficiency and not allowing dry growth, tuning thresholds for converting rimed snow to graupel, allowing graupel to sublimate outside of the cloud reducing

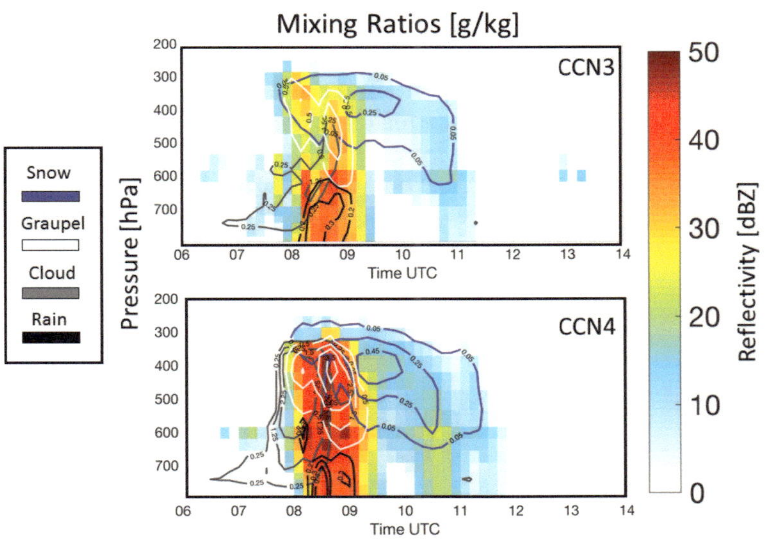

Figure 13.
*Temporal evolution of the mean vertical distribution of mixing ratio for snow, graupel, cloud and rain hydrometeors and reflectivity at the location defined by the orange box in **Figure 9** on May 15, 2009. Top—CCN3; bottom—CCN4. Contour interval is 1g/kg for cloud and graupel and 0.20g/kg for snow and rain starting at 0.25g/kg for the outermost contour line.*

amount of supercooled water available for riming, and by controlling the size of graupel and snow particles as a function of temperature [58]. However, the number of parameters involved is very high and the interaction of cold and warm microphysical processes with storm dynamics render this problem very difficult to address in a conclusive manner as many of the fundamental processes are not well understood yet [59–62]. The point made here is that when all else is the same, there is a significant change in precipitation patterns and 3D vertical structure depending on aerosol hygroscopicity, which is to say aerosol origin and source in the case of Central Nepal.

Analysis of the hydrometeor budgets suggests an emergent relationship between graupel production and CCN activation rate. This is illustrated by **Figure 14** that shows in the top row the vertical hydrometeor mixing ratio over the KWR for the remote aerosol (CCN3, left) and local aerosol (CCN4, right) during the time of heavier rainfall. In the bottom row, the vertical hydrometeor mixing ratio is shown for the IDR in the hour before (left panel) and during the storm (right panel) for the local aerosol. Both the KWR and the IDR areas (**Figure 2**) are on upwind slopes for the particular synoptic setup of the simulated storm: the KWR is on the southern facing slopes of the Himalayas on the upwind side with regard to westerly and southwesterly moisture advection at low levels, and the IDR is upwind with regard to the northwesterly storm propagation over the mountains. The figure shows that higher precipitation is associated with CCN4, and that the mixing ratio of graupel doubles when precipitation is heavier, and this behavior is robust across the region.

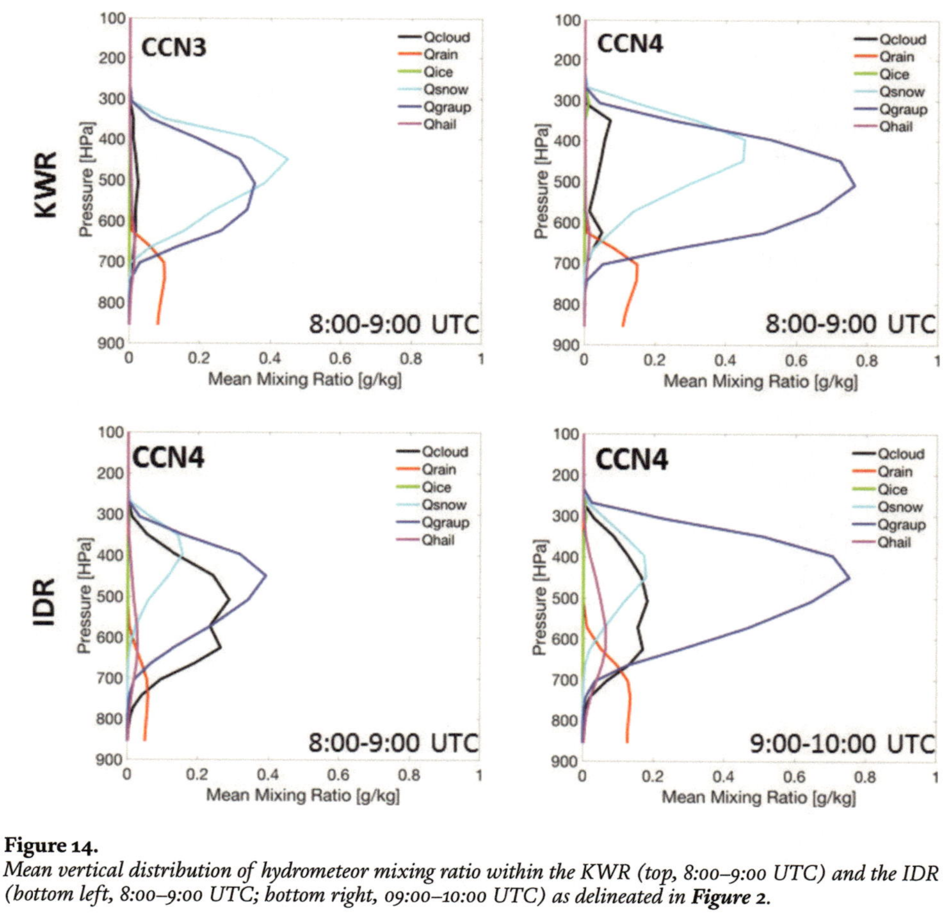

Figure 14.
Mean vertical distribution of hydrometeor mixing ratio within the KWR (top, 8:00–9:00 UTC) and the IDR (bottom left, 8:00–9:00 UTC; bottom right, 09:00–10:00 UTC) as delineated in ***Figure 2****.*

4. Summary and conclusions

A sensitivity study to examine the impact of CCN hygroscopicity on aerosol-cloud-precipitation interactions (ACPI) was conducted for a premonsoon storm in Central Nepal on May 15, 2009 during a major pollution event in the IGP using the WRF model as described in Section 2. CCN spectra used in this study included estimates from the in-situ measured aerosol size distribution and bulk hygroscopicity during JAMEX09 in Central Nepal [26, 37, 38]. Three distinct types of CCN activation spectra were used: the standard continental spectrum available in WRF, the CCN spectrum measured during the run-up of IGP pollution to the Kathmandu valley, here referred to as remote aerosol, and the CCN spectrum measured after the rainfall event when the remote aerosol was washed out and mostly aerosol from local sources is replenished in the atmosphere, here referred to as local aerosol.

An iterative method to estimate maximum supersaturation based on [46] was integrated to the double moment microphysics of MY05 [41] and implemented in WRF 8.3.1. The estimation of the maximum supersaturation in the code allows for a sensitivity study with CCN spectra fitted to the modified power law scheme of [45]. The results show that the differences in cumulative precipitation patterns between the standard and remote aerosols are small within 20%, but the differences between the simulations with remote and local aerosols are on the order of 25–50% and higher (**Figures** 5 and **6**). Interestingly, these large differences could be mapped for two catchments critical for hydropower (KWR) and water resources (IDR) in Central Nepal with much lower precipitation produced with the remote aerosol than with the local aerosol. The structure and spatial extent of the vertical wind component was observed to change among the simulations due to both microphysical and dynamic forcing, with topographic forcing playing an important role in the spatial organization of long-lived updrafts on upwind slopes at mid and high elevations.

Analysis of the space-time organization of precipitation, vertical winds and microphysics suggests that sustained graupel production is favored in long-lived updrafts (enhanced and maintained by terrain induced lifting and secondary convection) for slower CCN activation spectra (i.e., local aerosol). Higher graupel mixing ratios result in heavy localized rainfall. Because of role of the terrain in locking the spatial organization of vertical velocities depending on synoptic forcing, changes in CCN type as described by its activation spectra strongly impact the spatial patterns of rainfall accumulation at the ground.

This study illustrates the importance of specifying region-based CCN for modeling studies of aerosol-cloud-rainfall interactions and provides a first indication of the range of the uncertainty in the spatial variability of precipitation that can be attributed to aerosol sensitivity in the region. Nevertheless, detailed results from this simulation are specific for the specific storm, and in order to acquire a comprehensive understanding of regional aerosol-cloud-rainfall interactions, and the impact of IGP aerosol on rainfall in the Central Himalayas, more case studies need to be analyzed of climatologically relevant storm regimes and associated aerosol.

Acknowledgements

This research was supported by the Pratt School of Engineering.

Conflict of interest

The authors declare no conflict of interest.

Author details

Ana P. Barros*, Prabhakar Shrestha, Steven Chavez and Yajuan Duan
Duke University, Durham, North Carolina, USA

*Address all correspondence to: barros@duke.edu

References

[1] Barros AP. Orographic precipitation, freshwater resources, and climate vulnerabilities in mountainous regions. In: Climate Vulnerability: Understanding and Addressing Threats to Essential Resources. Vol. 5, Chapter 4. Elsevier: Elsevier Inc., 2013. pp. 57-78. DOI: 10.1016/B978-0-12-384703-4.00504-9

[2] Wilson AM, Barros AP. Landform controls on low level moisture convergence and the diurnal cycle of warm season orographic rainfall in the southern Appalachians. Journal of Hydrology. 2015;**531**(2):475-493. DOI: 10.1016/j.jhydrol.2015.10.068

[3] Wilson AM, Barros AP. Orographic land-atmosphere interactions and the diurnal cycle of low level clouds and fog. Journal of Hydrometeorology. 2017;**18**:1513-1533. DOI: 10.1175/JHM-D-16-0186.1

[4] Bollasina M, Nigam S. Absorbing aerosols and summer monsoon evolution over South Asia: An observational portrayal. Journal of Climate. 2008;**21**:3221-3239. DOI: 10.1175/2007JCLI2094.1

[5] Gautam R, Hsu NC, Lau KM, Tsay S-C, Kafatos M. Enhanced pre-monsoon warming over the Himalayan-Gangetic region from 1979-2007. Geophysical Research Letters. 2009;**36**:L07704. DOI: 10.1029/2009GL037641

[6] Gautam R, Hsu NC, Kafatos M, Tsay S-C. Aerosol and rainfall variability over the Indian monsoon region: Distributions, trends and coupling. Annales de Geophysique. 2009;**27**:3691-3703

[7] Shrestha P, Barros AP. Joint spatial variability of aerosol, clouds and rainfall in the Himalayas from satellite data. Atmospheric Chemistry and Physics. 2010;**10**:8305-8317. DOI: 10.5194/acp-10-8305-2010

[8] Lau WKM. The aerosol-monsoon climate system of Asia: A new paradigm. Journal of Meteorological Research. 2016;**30**(1):001-011. DOI: 10.1007/s13351-015-5999-1

[9] Rosenfeld D, Lohmann U, Raga GB, O' Dowd CD, Kulmala M, Fuzzi S, Reissell A, Andreae MO. Flood or drought: How do aerosols affect precipitation? Science. 2008;**321**:1309-1313. DOI: 10.1126/science.1160606

[10] Khain AP, Rosenfeld D, Pokrovsky A. Aerosol impact on the dynamics and microphysics of convective clouds Q. Journal of the Royal Meteorological Society. 2005;**131**:2639-2663. DOI: 10.1256/qj.04.62

[11] Seifert A, Beheng K. A two-moment cloud microphysics parameterization for mixed-phase clouds. Part II: Maritime versus continental deep convective storms. Meteorology and Atmospheric Physics. 2006;**92**:67-88

[12] Van den Heever SC, Carrico GG, Cotton WR, Demott PJ, Prenni AJ. Impacts of nucleating aerosol on Florida storms. Part I: Mesoscale simulations. Journal of the Atmospheric Sciences. 2006;**63**:1752-1775. DOI: 10.1175/JAS3713.1

[13] Tao W-K, Li X, Khain A, Matsui T, Lang S, Simpson J. Role of atmospheric aerosol concentration on deep convective precipitation: Cloud-resolving model simulations. Journal of Geophysical Research. 2007;**112**:D24S18. DOI: 10.1029/2007JD008728

[14] Lynn B, Khain A, Rosenfeld D, Woodley WL. Effects of aerosols on precipitation from orographic clouds. Journal of Geophysical Research. 2007;**112**(590):D10225. DOI: 10.1029/2006JD007537

[15] Fan J, Zhang R, Li G, Tao W-K. Effects of aerosols and relative humidity on cumulus clouds. Journal of Geophysical Research. 2007;**112**:D14204. DOI: 10.1029/2006JD008136

[16] Khain AP, Lynn B. Simulation of a supercell storm in clean and dirty atmosphere using weather research and forecast model with spectral bin microphysics. Journal of Geophysical Research. 2009;**114**:D19209. DOI: 10.1029/2009JD011827

[17] Khain AP. Notes on state-of-the-art investigations of aerosol effects on precipitation: A critical review. Environmental Research Letters. 2009;**4**:015004. DOI: 10.1088/1748-9326/4/1/015004

[18] Ntelekos AA, Smith JA, Donner L, Fast JD, Gustafson WI Jr, Chapman EG, Krajewski WF. The effects of aerosols on intense convective precipitation in the northeastern United States. Quarterly Journal of the Royal Meteorological Society. 2009;**135**:1367-1391

[19] Storer RL, Van Den Heever SC, Stephens GL. Modeling aerosol impacts on convective storms in different environments. Journal of the Atmospheric Sciences. 2010;**67**:3904-3915

[20] Duan Y, Petters MD, Barros AP. Understanding aerosol-cloud interactions in the development of orographic cumulus congestus during IPHEx. Atmospheric Chemistry and Physics Discussions. 2017. DOI: 10.5194/acp-2017-396

[21] Duan Y, Barros AP. Understanding how low-level clouds and fog modify the diurnal cycle of orographic precipitation using *in situ* and satellite observations. Remote Sensing. 2017;**9**(9):920, 33pp. DOI: 10.3390/rs9090920

[22] Duan Y. Mapping the Impact of Aerosol-Cloud Interactions on Cloud Formation and Warm-Season Rainfall in Mountainous Regions Using Observations and Models [Thesis]. Durham, North Carolina: Duke University; 2017

[23] Calvetti L, Filho AJP. Ensemble Hydrometeorological forecasts using WRF hourly QPF and top model for a middle watershed. Advances in Meteorology. 2014;**2014**:12. Article ID: 484120. DOI: 10.1155/2014/484120

[24] Radhakrishna B, Zawadzki I, Fabry F. Postprocessing model-predicted rainfall fields in the spectral domain using phase information from radar observations. Journal of the Atmospheric Sciences. 2013;**70**:1145-1159. DOI: 10.1175/JAS-D-12-0175.1

[25] Brun J, Shrestha P, Barros AP. Mapping aerosol intrusion in Himalayan valleys using the moderate resolution imaging Spectroradiometer (MODIS) and cloud–aerosol Lidar and infrared pathfinder satellite observation (CALIPSO). Atmospheric Environment. 2011;**45**:6382-6392. DOI: 10.1016/j.atmosenv.2011.08.026

[26] Shrestha P, Barros AP, Khlystov A. Chemical composition and aerosol size distribution of the middle mountain range in the Nepal Himalayas during the 2009 pre-monsoon season. Atmospheric Chemistry and Physics. 2010;**10**:11605-11621. DOI: 10.5194/acp-10-11605-2010

[27] Barros AP, Chiao S, Lang T, Burbank D, Putkonen J. From weather to climate—Seasonal and interannual variability of storms and implications for erosion processes in the Himalaya. Geological Society of America Special Papers. 2006;**398**:17-38. DOI: 10.1130/2006.2398 (02)

[28] Chiao S, Barros AP. A numerical study of the Hydrometeorological Dryline in Northwest India during the monsoon. Journal of the Meteorological Society of Japan. 2007;**85**:1-25

[29] Barros AP, Kim G, Williams E, Nesbit SW. Probing orographic controls in the Himalayas during the monsoon using satellite imagery. Natural Hazards and Earth System Sciences. 2004;**4**:1-23. DOI: 10.5194/nhess-4-29-2004

[30] Lang TJ, Barros AP. An investigation of the onsets of 1999 and 2000 monsoons in Central Nepal. Monthly Weather Review. 2002;**130**:1299-1316

[31] Pinty JP, Cosma S, Cohard J-M, Ricard E, Chaboreau J-P. CCN sensitivity of a warm precipitation event over fine scale orography with an advanced microphysical scheme. Atmospheric Research. 2001;**59-60**:419-446

[32] Saleeby SM, Cotton WR, Lowenthal D, Borys RD, Wetzel MA. Influence of cloud condensation nuclei on orographic snowfall. Journal of Applied Meteorology and Climatology. 2009;**48**:903-922. DOI: 10.1175/2008JAMC1989.1

[33] Muhlbauer A, Lohmann U. Sensitivity studies of the role of aerosols in warm-phase orographic precipitation in different dynamical flow regimes. Journal of the Atmospheric Sciences. 2008;**65**:2522-2542. DOI: 10.1175/2007JAS2492.1

[34] Iguchi T, Nakajimam T, Khain AP, Saito K, Takemura T, Suzuki K. Modeling the influence of aerosols on cloud microphysical properties in the East Asia region using a mesoscale model coupled with a bin-based cloud microphysics scheme. Journal of Geophysical Research. 2008;**113**:D14215. DOI: 10.1029/2007JD009774

[35] Lee SS, Donner LJ, Penner JE. Thunderstorm and stratocumulus: How does their contrasting morphology affect their interactions with aerosols? Atmospheric Chemistry and Physics. 2010;**10**:6819-6837. DOI: 10.5194/acp-10-6819-2010

[36] Dey S, Di Girolamo L. A climatology of aerosol optical and microphysical properties over the Indian subcontinent from 9 years (2000-2008) of multiangle imaging Spectroradiometer (MISR) data. Journal of Geophysical Research. 2010;**115**(D15):D15204. DOI: 10.1029/2009JD013395

[37] Shrestha P, Barros AP, Khlystov A. CCN estimates from bulk hygroscopic growth factors of ambient aerosols during the pre-monsoon season over Central Nepal. Atmospheric Environment. 2013;**67**:120-129. DOI: 10.1016/j.atmosenv.2012.10.042

[38] Shrestha P. Characterization of Pre-Monsoon Aerosol and Aerosol-Cloud-Rainfall Interactions in Central Nepal [Thesis]. Durham, North Carolina: Duke University; 2011

[39] Skamarock WC et al. A Description of the Advanced Research WRF Version 3. Boulder CO: National Center for Atmospheric Research; 2008; Report TN-475+STR. DOI: 10.5065/D68S4MVH

[40] Kain JS. The Kain–Fritsch convective parameterization: An update. Journal of Applied Meteorology. 2004;**43**:170-181

[41] Milbrandt JA, Yau MK. A multimoment bulk microphysics parameterization. Part II: A proposed three-moment closure and scheme description. Journal of the Atmospheric Sciences. 2005:3065-3081. DOI: 10.1175/JAS3816.1

[42] Iacono M et al. Radiative forcing by long-lived greenhouse gases: Calculations with the AER radiative transfer models. Journal of Geophysical Research. 2008;**113**:D13103. DOI: 10.1029/2008JD009944

[43] Tewari M et al. Implementation and verification of the unified NOAH land surface model in the WRF model. In: 20th conference on weather analysis

and forecasting/16th conference on numerical weather prediction; 2004. pp. 11-15

[44] Nakanishi M, Niino H. An improved Mellor–Yamada level-3 model with condensation physics: Its design and verification. Boundary-Layer Meteorology. 2004;**112**:1-31. DOI: 10.1023/B:BOUN.0000020164.04146.98

[45] Cohard J-M, Pinty J-P, Bedos C. Extending Twomey's analytical estimate of nucleated cloud droplet concentration from CCN spectra. Journal of the Atmospheric Sciences. 1998;**55**:3348-3357

[46] Cohard J-M, Pinty J-P. A comprehensive two-moment warm microphysical bulk scheme I: Description and tests. Quarterly Journal of the Royal Meteorological Society. 2000;**126**:1815-1842. DOI: 10.1002/qj.49712656613

[47] Diehl K, Simmel M, Wurzler S. Effects of drop freezing on microphysics of an ascending cloud parcel under biomass burning conditions. Atmospheric Environment. 2007;**41**:303-314. DOI: 10.1016/j.atmosenv.2006.08.011

[48] Hoose C, Kristjansson JE, Chen J-P, Hazra A. A classical-theory-based parameterization of heterogeneous ice nucleation by mineral dust, soot, and biological particles in a global climate model. Journal of the Atmospheric Sciences. 2010;**67**:2483-2503. DOI: 10.1175/2010JAS3425.1

[49] Medina S, Houze RA Jr. Air motions and precipitation growth in alpine storms. Quarterly Journal of the Royal Meteorological Society. 2003;**129**:345-337

[50] Barros AP, Joshi M, Putkonen J, Burbank DW. A study of the 1999 monsoon rainfall in a mountainous region in Central Nepal using TRMM products and raingauge observations. Geophysical Research Letters. 2000;**27**(22):3683-3686. DOI: 10.1029/2000GL011827

[51] Koren I, Remer LA, Altaratz O, Martins JV, Davidi A. Aerosol-induced changes of convective cloud anvils produce strong climate warming. Atmospheric Chemistry and Physics. 2010;**10**:5001-5010. DOI: 10.5194/acp-10-5001-2010

[52] Phillips VTJ, Donner LJ, Garner ST. Nucleation processes in deep convection simulated by a cloud-system-resolving model with double-moment bulk microphysics. Journal of the Atmospheric Sciences. 2007;**64**:738-761. DOI: 10.1175/JAS3869.1

[53] Lim K-SS, Hong S-Y, Yum SS, Dudhia J, Klemp JB. Aerosol effects on the development of a supercell storm in a double-moment bulk-cloud microphysics scheme. Journal of Geophysical Research. 2011;**116**:d02204. DOI: 10.1029/2010JD014128

[54] Lascaux F, Richard E, Pinty J-P. Numerical simulation of three different MAP IOPs and the associated microphysical processes. Quarterly Journal of the Royal Meteorological Society. 2006;**132**:1907-1926. DOI: 10.1256/qj.05.197

[55] Yuter SE, Houze RA Jr. Three-dimensional kinematic and microphysical evolution of Florida cumulonimbus. Part II: Frequency distributions of vertical velocity, reflectivity, and differential reflectivity. Monthly Weather Review. 1995;**123**:1941-1963

[56] Lang S, Tao W-K, Simpson J, Ferrier B. Modeling of convective-Stratiform precipitation processes: Sensitivity to partitioning methods. Journal of Applied Meteorology. 2003;**42**:505-527

[57] Lang SE, Tao W-K, Zeng X, Li Y. Reducing the biases in simulated radar

Reflectivities from a bulk microphysics scheme tropical convective systems. Journal of the Atmospheric Sciences. 2011;**68**:2306-2320. DOI: 10.1175/JAS-D-10-05000.1

[58] Tao W-K et al. The impact of microphysical schemes on hurricane intensity and track. Asia-Pacific Journal of Atmospheric Sciences. 2011;**47**(1): 1-16. DOI: 10.1007/s13143-011-1001-z

[59] Tao W-K et al. High-resolution NU-WRF model simulations of MC3E, deep convective-precipitation systems: Comparisons between Goddard microphysics schemes and observations. Journal of Geophysical Research – Atmospheres. 2016;**121**:1278-1306. DOI: 10.1002/2015JD023986

[60] Fan J et al. Cloud-resolving model intercomparison of an MC3E squall line case: Part I—Convective updrafts. Journal of Geophysical Research – Atmospheres. 2017;**122**:9351-9378. DOI: 10.1002/2017JD026622

[61] Barnes HC, Houze R Jr. Comparison of observed and simulated spatial patterns of ice microphysical processes in tropical oceanic mesoscale convective systems. Journal of Geophysical Research-Atmospheres. 2016;**121**: 8269-8296. DOI: 10.1002/2016JD025074

[62] Khain AP et al. Representation of microphysical processes in cloud resolving models: Spectral (bin) microphysics versus bulk parameterization, rev. Geophysics. 2015;**53**:247-322. DOI: 10.1002/2014RG000468

Chapter 6

The Chemical Nature of Individual Size-Resolved Raindrops and Their Residual Particles Collected During High Atmospheric Loading for $PM_{2.5}$

Chang-Jin Ma and Gong-Unn Kang

Abstract

Although it is well known that rain plays an important role in capturing air pollutants, its quantitative evaluation has not been done enough. In this study, the pollutant scavenging effect by size of raindrops was investigated by clarifying the chemical nature of individual size-resolved raindrops and their residual particles. Raindrops as a function of their size were collected using the raindrop collector devised by ourselves during high atmospheric loading for $PM_{2.5}$. The raindrop number concentration (m^{-2} h^{-1}) tended to drastically decrease as the drop size goes up. Particle scavenging rate, $R_{sca.}$(%), based on the actual measurement values were 38.7, 69.5, and 80.8% for the particles with 0.3–0.5, 0.5–1.0, and 1.0–2.0 μm diameter, respectively. S, Ca, Si, and Al ranked relatively high concentration in raindrops, especially small ones. Most of the element showed a continuous decrease in concentration with increasing raindrop diameter. The source profile by factor analysis for the components of residual particles indicated that the rainfall plays a valuable role in scavenging natural as well as artificial particles from the dirty atmosphere.

Keywords: raindrop, washout, $PM_{2.5}$, scavenging, PIXE

1. Introduction

Fine particulate matter smaller than 2.5 μm ($PM_{2.5}$) is already a serious and growing environmental problem in China. The smog situation in China is getting even worse. Extremely high levels (several tens of times higher than the concentration of WHO's recommendation (<25 μg m^{-3} of 24-hour exposure) can last more than a week.

Now, the damage caused by $PM_{2.5}$ is not just an issue for China. It has been proven that the worsening of air pollution in China affects Korea and Japan as well, with the fine particles moving with the winds. Especially, air quality in South Korea has become an increasing source of concern for the nation. According to the survey conducted by the Seoul Development Institute [1], 52% of the residents in Seoul consider air pollution to be the most imminent environmental issue, and 68% of them consider that the level of air pollution is serious.

In the case of Japan, in recent years, high levels of $PM_{2.5}$ were recorded in parts of western Japan, especially Fukuoka Prefecture [2]. Under the growing concern to this pollution crisis, many regional organizations in Japan are planning various measures like providing the real-time hourly $PM_{2.5}$ data on website.

This fiendish $PM_{2.5}$ are removed from the atmosphere by wet and/or dry depositions, the former proceeds more efficiently in the form of precipitation such as rain, fog, and snow [3–5]. Most people have probably experienced a clear view of the sky and landscapes after rainfall. This is a good example to understand that the rain is a great cleaner of $PM_{2.5}$. Therefore, in the assessment of public health risks associated with $PM_{2.5}$, a study of the chemical nature of the rain have fallen during high atmospheric loading for $PM_{2.5}$ is no less important than that of the ambient $PM_{2.5}$ itself.

One of major mechanisms of the incorporation of ambient particles into raindrops is the collision among the particles below the cloud base. The efficiency of the collision depends on the size distributions of particles and the raindrops [3, 5]. Therefore, the exposure amount of pollutants by rain is variable depend on rainfall properties (e.g., rainfall amount, rainfall duration, rainfall intensity, and raindrop size distribution). More than all, as mentioned above, raindrop size distribution is crucially important because it plays an important role in capturing pollutants.

The collection of raindrops as a function of their size has not been generalized because it is technically difficult to capture and a high degree of skill in handling of raindrops is demanded [6, 7].

This study has been carried out to improve our understanding of particle scavenging properties of size-resolved raindrops from the chemical properties of the residuals in individual raindrops.

2. Experimental methods

2.1 Collection and handling of size-resolved raindrops

For the sampling of single raindrops as a function of their size, the raindrop collector devised by our oneself in previous research [8] was applied. It was located on the rooftop of a four-story building (a height of 20 m above ground level) (33.40 N 130.26 E) at Fukuoka Women's University in Fukuoka City, Japan at a time when it was raining. The rain event lasted 14 h, with rain intensity of 0.2–1.2 mm h^{-1}, and air temperature between 9.6 and 16.4°C. The concentration of $PM_{2.5}$ in that time was higher than the Japanese central government's safety standard for $PM_{2.5}$ (i.e., a mean of 35 μg m^{-3} over a 24-hour period).

Although more details can be found in our previous papers [6, 8, 9], the principle of our own raindrop capture device can be summarized as follows:

Fallen raindrops into the liquid nitrogen are frozen and they sink to lower sieves owing to their higher density as illustrated at the top in **Figure 1**. As shown in **Figure 2**, the raindrops kept their spherical shape during the freezing process, and consequently by using stainless steel sieves of different mesh widths (1.7 mm, 0.17 mm, and back-up) it is possible to separate the frozen raindrops according to their sizes.

Our previous study [8] has been clearly presented that there was no meaningful size change of raindrop when it was freezing. In other words, it can be said that diameter change of frozen raindrops by liquid nitrogen did not affect the size segregation of our raindrop collector.

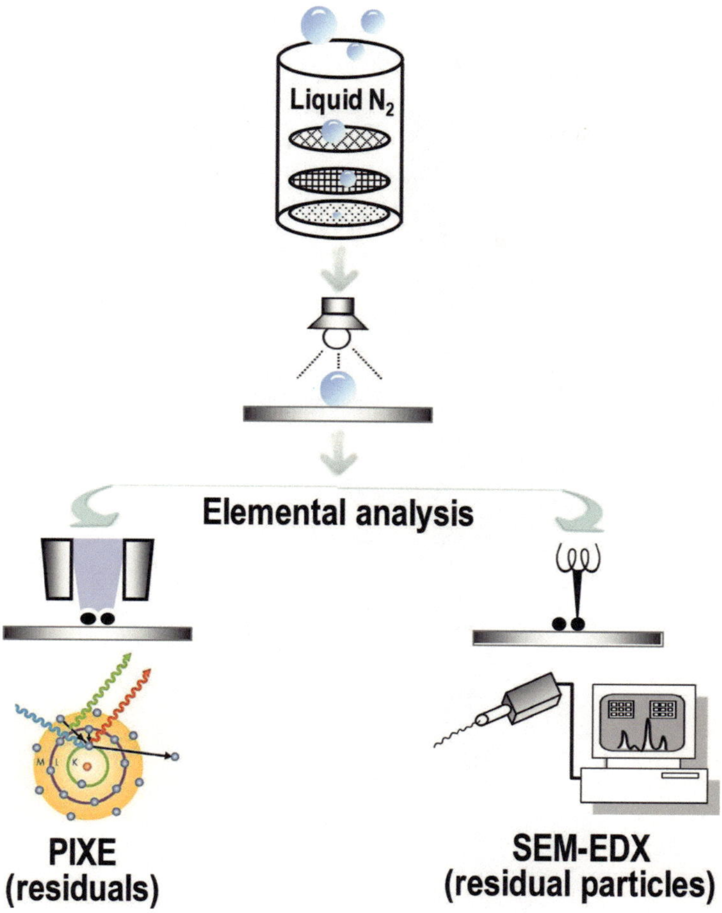

Figure 1.
Flow of collection, pretreatment, and analysis of individual size-resolved raindrops and their residual particles.

After sampling, the sieves were pulled out from the dewar vacuum flask and each frozen raindrop on each sieve was placed onto the non-hole Nucleopore® filter and Ag thin film (99.99% purity) by using a vacuum pipette (HAKO 392). The frozen raindrops were melted and dried under an infrared lamp for 5 min. Because every process was performed in the clean air system filled with the cooling nitrogen gas, the raindrop handling could be done successfully without any loss of some residues, evaporation, and contamination.

Moreover, in order to measure the rainfall intensity, the standard rain gauge (260-2510, NovaLynx Co.) consisting of a funnel that empties into a graduated cylinder was applied.

2.2 Elemental analyses of residues in raindrops

Elemental analyses of solid residues and individual residual particles in raindrops were subsequently analyzed by particle induced X-ray emission (PIXE) and scanning electron microscopy (SEM) with energy dispersive X-ray analysis (EDX), respectively. The PIXE installed at the Cyclotron Research Center of Iwate Medical University was applied and it has the great advantages such as an excellent sensitivity, a nondestructive technique for multielement with a wide range of elements ($Z > 10$). The sensitivity, if defined by the ratio between PIXE yield per unit dose and mass thickness, can be determined for all objective elements both

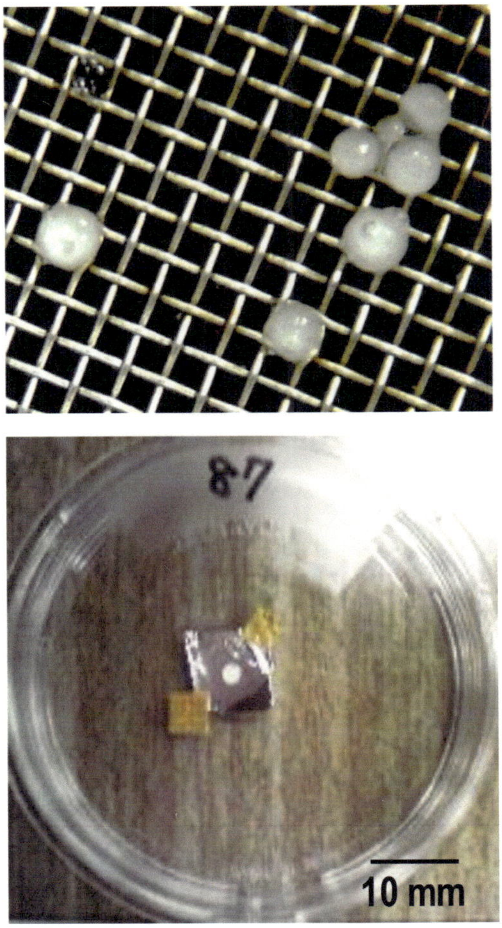

Figure 2.
Several frozen raindrops lying on the sieve with 1.7 mm mesh size (top) and a raindrop on Ag film (bottom).

experimentally and theoretically. For instance, the sensitivity of calcium was calculated to be 1700 (counts $cm^2/\mu C\ \mu g$) with a detection limit of 9.4×10^{-3} ($\mu g/cm^2$). The more detailed analytical procedures and experimental setup for PIXE analysis were described elsewhere [10].

The overall process of collection and handling of size-resolved raindrops, and their elemental analyses is shown in **Figure 1**.

2.3 Monitoring of sulfate, $PM_{2.5}$, and particle number concentration

In addition to the raindrop collection, the highly time-resolved (10-minute cycle) $PM_{2.5}$ sulfate was monitored during a whole research period by the ambient particulate sulfate monitor (8400S, Rupprecht & Patashnick Co.). A detailed description of design, operation, and data reduction and processing can be found elsewhere [11]. In order to measure $PM_{2.5}$ mass concentration, a light scattering $PM_{2.5}$ monitors (Dust Scan Scouts 3020, Rupprecht & Patashnick Co.) was simultaneously operated. Details on this $PM_{2.5}$ monitoring system was previously described [12].

The number concentrations of size-selective particulate matters (i.e., 0.3–0.5, 0.5–1.0, 1.0–2.0, and 2.0–5.0 μm) were also monitored by an optical particle counters (OPC) (RION, KC-01D).

3. Results and discussion

3.1 Number size distribution of raindrops

The number size distribution of raindrops collected at two different rainfall intensities was estimated by the volume of melted raindrops and that of calculated single raindrop from its average size. **Figure 3** shows the raindrop number size distribution at the beginning (0.2 mm h^{-1}) and subsequent (1.2 mm h^{-1}) rainfalls. The raindrop concentration Nr (m^{-2} h^{-1}) is the total number of drops falling per square meter of surface per hour. Raindrop number tended to drastically decrease as the drop size goes up at both hourly rain rates (i.e., rain intensity). Needless to say, it showed higher number concentration when the higher rainfall intensity. This result indicates that the increase in the rain rate stemmed mainly from the increase in the number concentration of raindrops with drop diameter < 0.94 mm.

3.2 Calculation of particle scavenging efficiency of size-resolved raindrops

Prior to the interpretation of the actual measurement data about particle scavenging properties of size-resolved rain drops, the collection efficiency of ambient particles as a function of raindrop size was theoretically calculated.

Slinn and Hales [13] proposed three particle collection efficiencies (E), that is, E by Brownian diffusion ($E_{dif.}$), E by interception ($E_{int.}$), and E by inertial impaction processes ($E_{imp.}$). Subsequently, Strauss [14] suggested a more advanced equation for particle collection efficiency ($E_{integ.}$) of raindrops that integrated three kinds of efficiencies as follows:

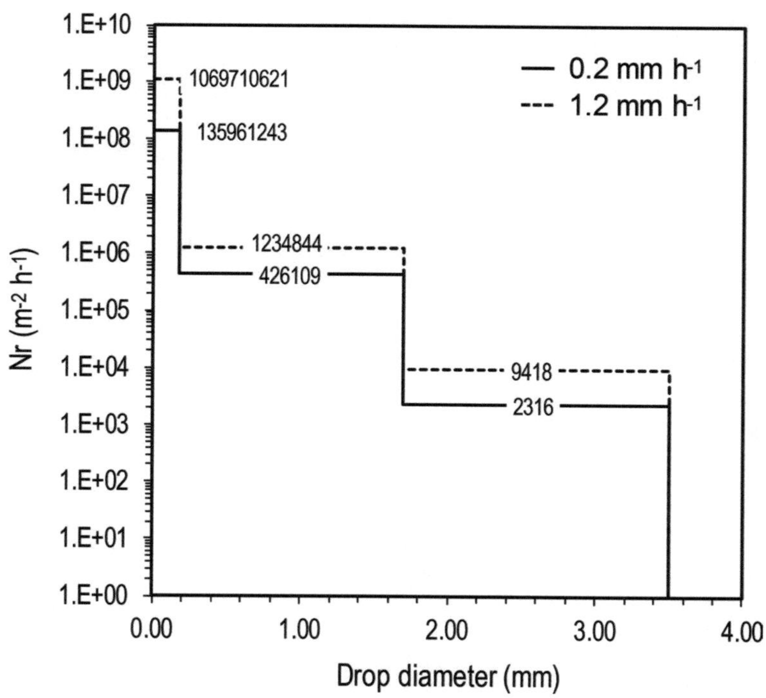

Figure 3.
Raindrop number size distribution at the beginning (0.2 mm h^{-1}) and subsequent (1.2 mm h^{-1}) rainfalls.

$$E_{integ.} = 1 - (1 - E_{dif.})\ (1 - E_{int.})\ (1 - E_{imp.}) \quad (1)$$

In this study, the collection efficiency of ambient particles for three kinds of raindrop sizes (i.e., 0.09, 0.94, and 2.60 mm diameter) was theoretically computed using this integrated $E_{integ.}$ Details for calculation like variable settings and computation processes were already described in our earlier articles [6, 15]. **Figure 4** shows the below-cloud scavenging coefficients of ambient particles by falling raindrops as a function of aerosol radius and collector rain drop size. There are several features that appear in the particle capturing efficiency curves. As the first outstanding feature, the smaller raindrop can scavenge particles more efficiently. Another peculiarity for particle scavenging efficiency curves is that the efficiency is rapidly degraded in the central part of each curve. This is well-known phenomenon as the term of "Greenfield gap" [13]. Although, the margin is varied from 0.01 to 2 μm [16], this gap has been reported in various studies.

3.3 Particle scavenging properties of size-resolved raindrops

Figure 5 shows time series variation of relative humidity and four-size resolved (0.3–0.5, 0.5–1.0, 1.0–2.0, and 2.0–5.0 μm) particle number concentration throughout a whole rainfall duration. As previously mentioned, it was a relatively weak rainfall with the rainfall intensity varied from 0.2 to 1.2 mm h^{-1}.

Particles of all sizes exhibit lower concentrations when the rain started. Although this was the expected result, an unusual thing was that particle number concentration was temporarily increased immediately after the rain.

Decreasing relative humidity increases scavenging in the Greenfield gap since the evaporating raindrops are cooler at the surface, and this sets up a thermal gradient that induces motion of the aerosols towards the cooler raindrop surface [17]. In the condition of initial rainfall, falling raindrops can be exposed to the dry air of near the surface. According to Croft et al. [17], this situation will reduce the levels

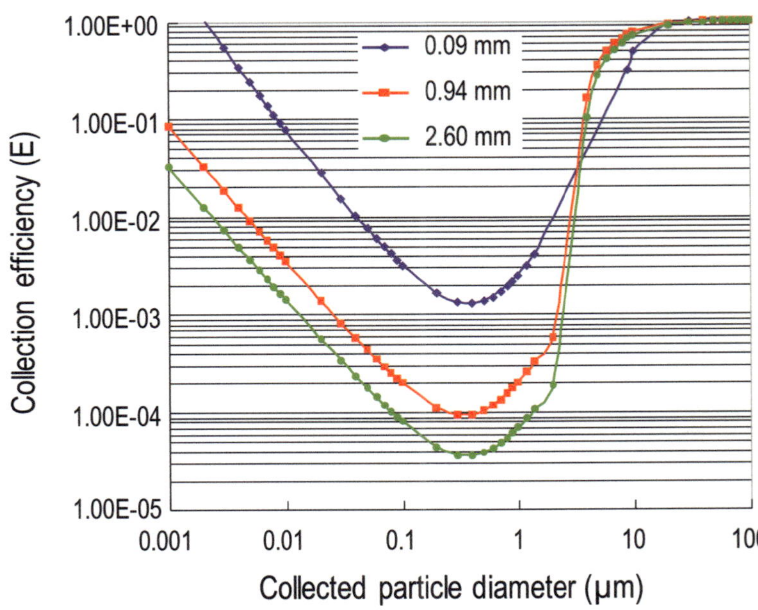

Figure 4.
Theoretically calculated collection efficiency of ambient particles as a function of raindrop size.

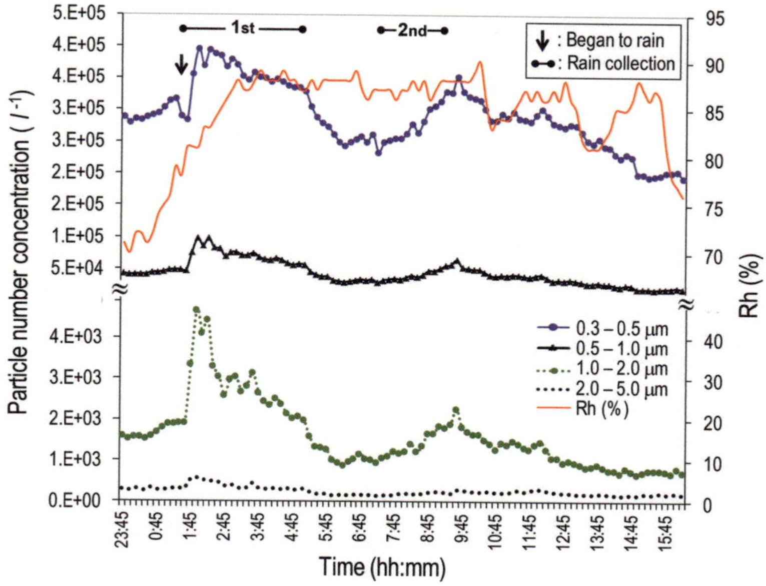

Figure 5.
Time series variation of relative humidity and size-resolved particle number concentration.

of atmospheric particles because the particle removal efficiency can be improved by increasing scavenging in the Greenfield gap. Nevertheless, the reason for a temporarily increasing of particle concentration might be that new particles were created in the early stage of precipitation because small size rain droplets cloud be easily evaporated in the dried low altitude air.

A continuous decrease of the particle number concentration over most of the range of particle diameter during a whole precipitation was not seen. There was a modest increase in particle number concentration from 07:00 to 09:30 (see **Figure 5**). Such increases might be caused by a decline of rainfall intensity and the influx of pollutants to our measurement site.

Scavenging rate, $R_{sca.}$(%), of three-size resolved (0.3–0.5, 0.5–1.0, and 1.0–2.0 μm) particles showing an obvious reduction was calculated by below equation:

$$R_{sca.}(\%) = \frac{N_{p_{max.}} - N_{p_{min.}}}{N_{p_{max.}}} \times 100 \qquad (2)$$

where $N_{pmax.}$ and $N_{pmin.}$ are the maximum and minimum particle number concentration for target particle size, respectably.

Calculated $R_{sca.}$ based on the actual measurement values were 38.7, 69.5, and 80.8% for the particles with 0.3–0.5, 0.5–1.0, and 1.0–2.0 μm diameter, respectively. The results show a good match to that of model calculation shown in **Figure 4**.

Figure 6 shows the time series variation of $PM_{2.5}$, sulfate in $PM_{2.5}$, their $R_{sca.}$, and relative humidity in the first half of rainfall. Here, let us pay attention to the $R_{sca.}$ for $PM_{2.5}$ and sulfate in $PM_{2.5}$. $R_{sca.}$ shows a gradual increase in both. However, that of $PM_{2.5}$ (63.1%) was overwhelmingly high compared to sulfate in $PM_{2.5}$ (33.0%).

Despite sulfate particles being water soluble, unexpectedly they have a low $R_{sca.}$ Most of the sulfate in fine particles are forming $(NH_4)_2SO_4$ or NH_4HSO_4 and they are generally appearing in the fine mode at 0.5–0.6 μm [18]. Therefore, Greenfield gap is the reason for the low $R_{sca.}$ of sulfate in $PM_{2.5}$.

Meanwhile, there was a very high rain washing efficiency for $PM_{2.5}$. This result should be welcomed by people, particularly living in East Asia, who suffer from $PM_{2.5}$.

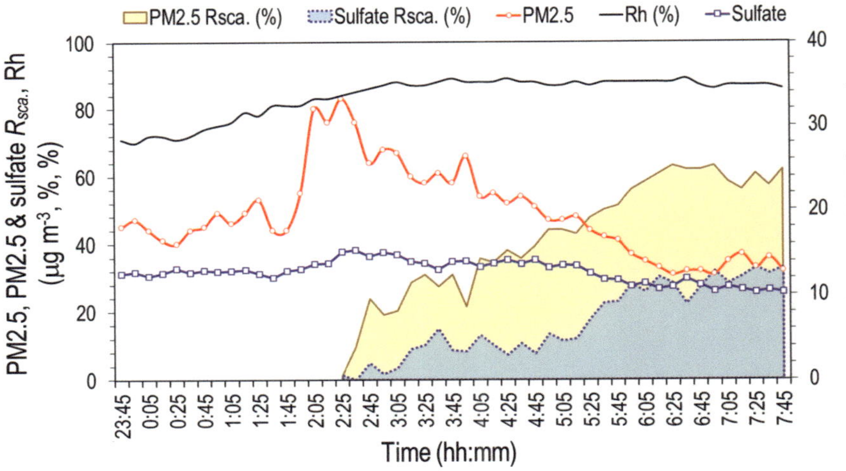

Figure 6.
Time series variation of PM2.5 ($\mu g\ m^{-3}$), sulfate in PM2.5 ($\mu g\ m^{-3}$), their Rsca. (%), and relative humidity (%).

3.4 Chemical properties of the residuals in size-classified raindrops

Figure 7 shows the elemental concentration of individual raindrops classified in three steps (i.e., 0.09, 0.94, and 2.6 mm) in order of size. The data were the elemental concentration in the raindrops at the beginning (0.2 mm h^{-1}) rainfall and they were determined by PIXE analysis. S, Ca, Si, and Al ranked relatively high concentration in raindrops, especially small ones. Most of the element showed a continuous decrease in concentration with increasing raindrop diameter. Especially, there was a marked decrease in the range of between 0.09 and 0.94 mm raindrop diameter. Although little is known at present, it is expected that smaller raindrops should have higher elemental concentration because they have lower falling velocities and consequently they can effectively remove pollutants during longer lifetimes than larger ones. On the other hand, in some of the elements (e.g., Cl, Ca, Cu, and Zn), a slight increase was found between 0.94 and 2.6 mm raindrop diameter. If the elemental constituent were originated from large size particle (>4 mm in diameter), it has some possibility of the increasing for elemental concentration between 0.94 and 2.6 mm raindrops in diameter. Among many elements determined from individual raindrops, sulfur that showed the highest concentration was probably derived from gaseous SO_2 and particulate sulfur taken up by falling raindrops. As another remarkable thing, unexpectedly, relatively high levels of lead were detected. Their concentrations were 0.83, 0.28, and 0.088 ppm in 0.09, 0.94, and 2.6 mm raindrops, respectively. In dealing with the health hazards of lead like neurologic and behavioral disorders, it is something that we ought to be talking about. In general, major sources of ambient lead are piston-engine aircraft operating on leaded aviation fuel, metals processing, waste incinerators, and lead-acid battery manufacturers. Although, because of its serious impacts on public health, leaded gasoline was permanently banned in Japan long time ago, the metal contaminations of roadside soils were included in ambient particles until now [19, 20].

3.5 Chemical properties of the residual particles in size-classified raindrops

When melt and evaporation of frozen raindrops were completed, it was possible to maintain residual materials on the Ag thin film. These retained matters could be the targets of SEM–EDX analysis. An example of EDX spectrum and elemental Weight% of a single residual particle in a raindrop (D_r = 2.5 mm) was drawn in **Figure 8**.

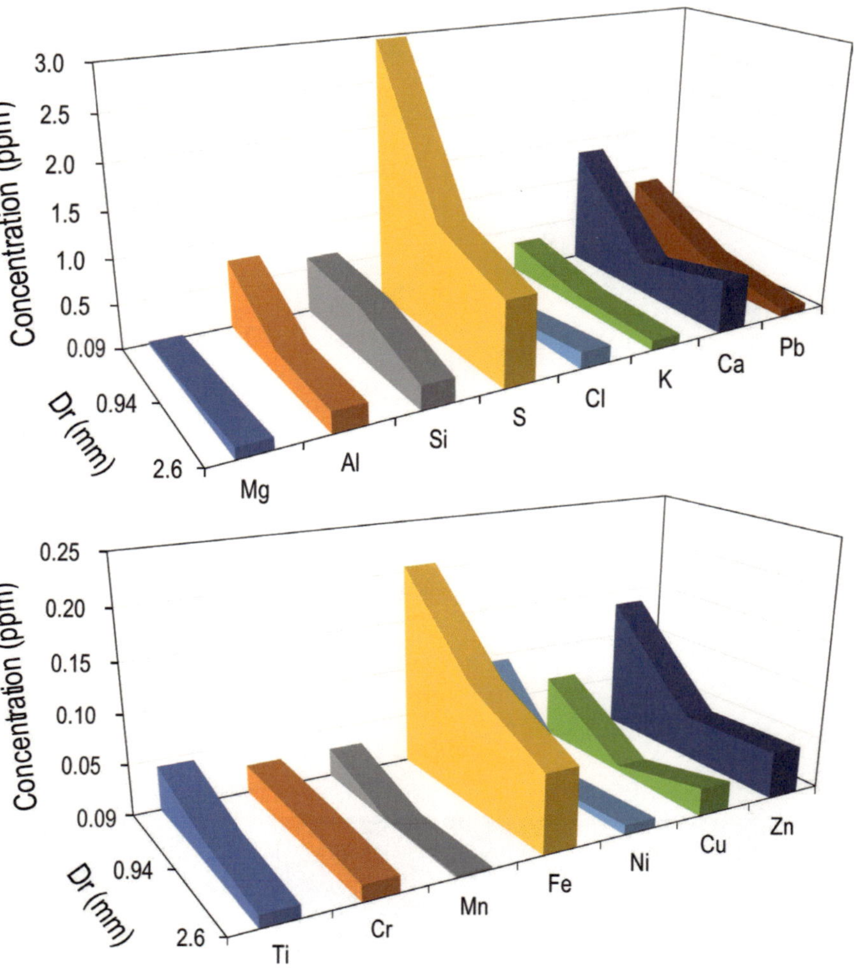

Figure 7.
Elemental concentration of size-resolved raindrops determined by PIXE analysis.

Figure 9 illustrates the SEM image (top left) of a single residual particle in a raindrop (D_r = 2.5 mm) and its elemental maps. These visualized elemental maps for several elements including oxygen enable us not only to presume the chemical mixing state of raindrop residual particles, but also to estimate their source profiles.

To obtain the source profile information for aerosol components, factor analysis, which is one of the multivariate statistical techniques, was carried out. It was possible to construct the matrix of 450-set of 13 components by the result of SEM–EDX analysis for residual particles in a total of 150 individual raindrops. The factor loadings (Varimax with Kaiser normalization) are shown in **Table 1**. The data matrix (450 variables × 13 cases) constructed in the present study was successfully classified into four factors. The first factor (34% of the variance) shows high loadings for S, P, Cu, Ca, and Cr. Although typical soil component like Al and Si were excluded because of their missing data, the first factor was associated with soil components (Ca and other minors). The second factor, which explains 26% of the variance, seemed to express the sources of fossil fuel combustion. The third factor grouped F and Fe were the components originated from biomass burning (or volcanoes) [21] and iron industry. The fourth factor dominantly made up of typical marine components. The cumulative variance of these three factors was 85.2%. This

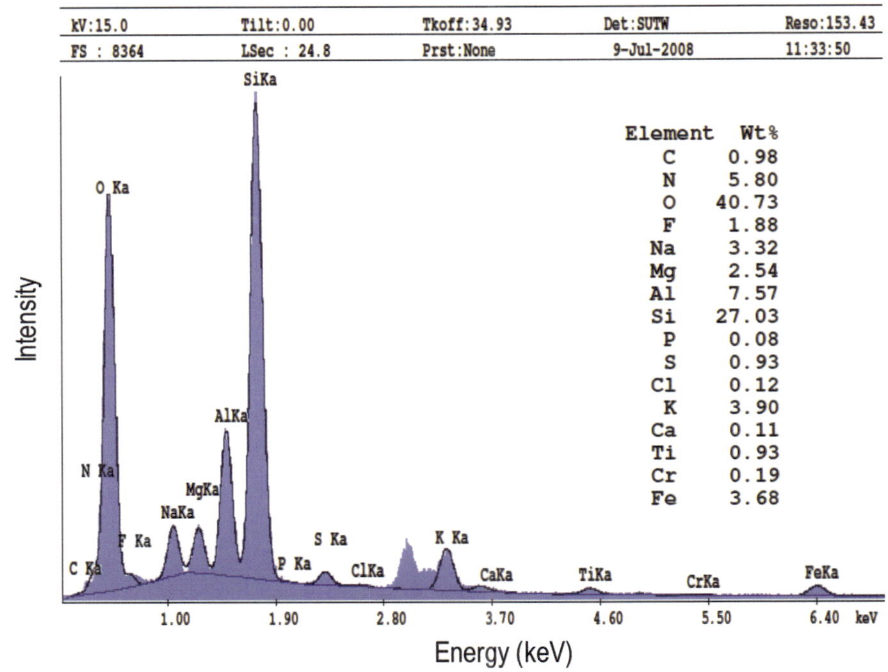

Figure 8.
EDX spectrum and elemental weight% of a single residual particle in a raindrop (D_r = 2.5 mm).

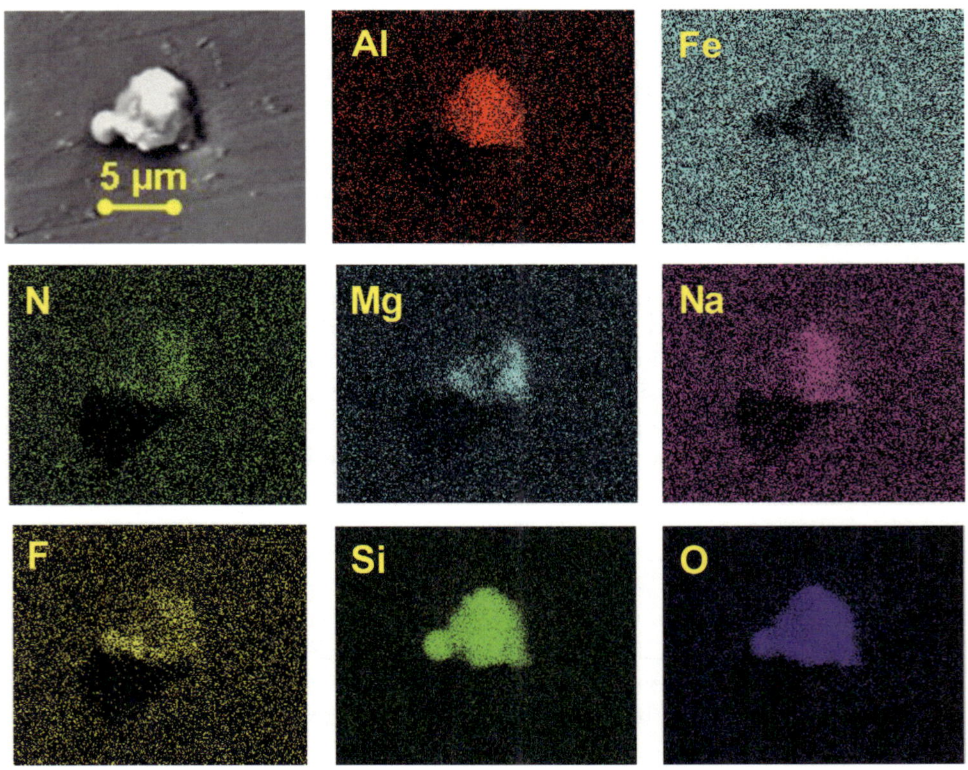

Figure 9.
SEM image (top left) of a single residual particle in a raindrop (D_r = 2.5 mm) and its elemental maps

	Factor 1	Factor 2	Factor 3	Factor 4
S	0.901	0.128	-0.030	-0.026
P	0.853	0.113	0.011	-0.060
Cu	0.780	0.057	-0.052	0.033
Ca	0.756	-0.039	0.084	0.148
Cr	0.565	0.502	0.042	-0.037
C	-0.056	0.976	-0.059	-0.012
N	0.430	0.786	-0.110	0.303
F	0.037	-0.015	0.960	0.101
Fe	-0.098	-0.078	0.936	-0.234
K	0.177	0.096	-0.144	0.805
Cl	0.467	0.425	0.064	0.688
Mg	-0.200	-0.039	0.140	0.597
Na	-0.141	0.168	-0.198	0.493

Extraction method: Principal component analysis
Rotation method: Varimax with Kaiser normalization

Table 1.
Result of factor analysis for the elements detected from individual residues in raindrops with diameters between 1.17 and 3.5 mm.

result indicates that the rainfall plays a valuable role in scavenging natural as well as artificial particles from the dirty atmosphere.

4. Conclusions

Ambient particles are ultimately removed from the atmosphere by the natural processes generally referred to as wet precipitation and dry deposition. The former is the most important natural removal mechanism of ambient air pollutants including $PM_{2.5}$. The clear landscapes after the rain is a good proof of this wet precipitation. In this study, the scavenging properties of ambient particles were investigated by collection of raindrops as a function of their size and clarifying their chemical nature. Particle scavenging rates based on both the actual measurement and the theoretically calculated values indicated that raindrops, especially small raindrops, played a great role to remove ambient particles including both naturally and artificially formed ones. On the other hand, through the PIXE analytical technique for the residuals in and/or on individual raindrops, it was obvious that several hazard components like Pb, Cr, and Mn had meaningful amount. Therefore, we must be thankful for the role of the rain, at the same time, we should avoid getting wet to protect our health from heavy metals. Further study on a comparison of the real measured data with the calculated result for the particle scavenging by raindrops is being planned in the near future.

Author details

Chang-Jin Ma[1*] and Gong-Unn Kang[2]

1 Department of Environmental Science, Fukuoka Women's University, Fukuoka, Japan

2 Department of Medical Administration, Wonkwang Health Science University, Iksan, Korea

*Address all correspondence to: ma@fwu.ac.jp

References

[1] Chung A. Korea's policy towards pollution and fine particle: A sense of urgency. Korea Analysis No. 2. 2014. Available from: http://www.centreasia.eu/sites/default/files/publications_pdf [Accessed: 2017-01-15]

[2] Fukuoka issues warning on high PM2.5 levels. The Japan Times. 2013. Available from: http://www.japantimes.co.jp/news [Accessed: 2017-01-15]

[3] Byrne MA, Jennings SG. Scavenging of submicrometer aerosol particles by water drops. Atmospheric Environment. 1993;**27A**:2099-2105

[4] McGann K, Haag I, Roder A. The efficiency with drizzle and precipitation sized drops collide with aerosol particles. Atmospheric Environment. 1993;**25A**:791-799

[5] Flossmann AI, Hall WD, Purppacher HR. A theoretical study of the wet removal of atmospheric pollutants. Part I: The redistribution of aerosol particles captured through nucleation and impaction scavenging by growing cloud drops. Journal of the Atmospheric Sciences. 1985;**42**:583-606

[6] Ma CJ, Tohno S, Kasahara M, Hayakawa S. The nature of individual solid particles retained in size-resolved raindrops fallen in Asian dust storm event during ACE-Asia, Japan. Atmospheric Environment. 2004;**38**:2951-2964

[7] Tenberken B, Bächmann K. Analysis of individual raindrops by capillary zone electrophoresis. Journal of Chromatography A. 1996;**755**:121-126

[8] Ma CJ, Kasahara M, Hwang KC, Choi KC, Kim HK. Measurement of the single and size-classified raindrops. Journal of Korea Society for Atmospheric Environment. 2000;**15E**:73-78

[9] Ma CJ, Kasahara M, Tohno S. A new approach for characterization of single raindrops. Water, Air and Soil Pollution. 2001;**130**:1601-1606

[10] Sera K, Futatsugawa S, Matsuda K. Quantitative analysis of untreated bio-samples. Nuclear Instruments and Methods in Physics Research B. 1999;**150**:226-233

[11] Drewnick F, Schwab JJ, Hogrefe O, Peters S, Diamond D, Weber R, et al. Intercomparision and evaluation of four semi-continuous PM2.5 sulfate instruments. Atmospheric Environment. 2003;**37**:3335-3350

[12] Ma CJ, Kim KH. Artificial and biological particles in the springtime atmosphere. Asian Journal of Atmospheric Environment. 2013;**7**:209-216

[13] Slinn WGN, Hales JM. A reevaluation of the role of thermophoresis as a mechanism of in- and below-cloud scavenging. Journal of Atmospheric Science. 1971;**28**:1465-1471

[14] Strauss W. Industrial Gas Cleaning. New York: Pergamon Press; 1975. pp. 293-3005

[15] Ma CJ, Tohno S, Kasahara M. Experimental and model studies on the size-resolved raindrops collected in the episodical yellow rainfall event. Journal of Aerosol Research. 2005;**20**:324-332

[16] Henzing JS, Olivié DJL, Van Velthoven PFJ. A parameterization of size resolved below cloud scavenging of aerosols by rain. Atmospheric Chemistry and Physics. 2006;**6**:3363-3375

[17] Croft B, Lohmann U, Martin RV, Stier P, Wurzler S, Feichter J, et al. Aerosol size-dependent below-cloud scavenging by rain and snow in

the ECHAM5-HAM. Atmospheric Chemistry and Physics. 2009;**9**:4653-4675

[18] Taiwo AM, Beddows DCS, Shi Z, Harrison RM. Mass and number size distributions of particulate matter components: Comparison of an industrial site and an urban background site. Science of the Total Environment. 2014;**475**:29-38

[19] Chen X, Xia X, Zhao Y, Zhang P. Heavy metal concentrations in roadside soils and correlation with urban traffic in Beijing, China. Journal of Hazardous Materials. 2010;**181**:640-646

[20] Christoforidis A, Stamatis N. Heavy metal contamination in street dust and roadside soil along the major national road in Kavala's region, Greece. Geoderma. 2009;**151**:257-263

[21] Jayarathne T, Stockwell CE, Yokelson RJ, Nakao S, Stone EA. Emissions of fine particle fluoride from biomass burning. Environmental Science & Technology. 2014;**48**:12636-12644